# Love Online

# Love Online

---

# Jean-Claude Kaufmann

## Translated by David Macey

polity

First published in French as *Sex@mour* by Armand Colin, 2010

The English edition © Polity Press, 2012

Polity Press
65 Bridge Street
Cambridge CB2 1UR, UK

Polity Press
350 Main Street
Malden, MA 02148, USA

ISBN-13: 978-0-7456-5183-5
ISBN-13: 978-0-7456-5184-2(pb)

A catalogue record for this book is available from the British Library.

Typeset in 11 on 13 pt Sabon
by Toppan Best-set Premedia Limited
Printed and bound in Great Britain by the MPG Books Group Limited

The publisher has used its best endeavours to ensure that the URLs for external websites referred to in this book are correct and active at the time of going to press. However, the publisher has no responsibility for the websites and can make no guarantee that a site will remain live or that the content is or will remain appropriate.

Every effort has been made to trace all copyright holders, but if any have been inadvertently overlooked the publisher will be pleased to include any necessary credits in any subsequent reprint or edition.

For further information on Polity, visit our website: www.politybooks.com

*David Macey 1949–2011*

A translator does not just translate; he is really co-author of any new book that passes through his fingers. David Macey was an exceptional co-author, at once respectful of the original text and inventive and careful in its translation. His death leaves a great void. But his books will survive him. This one is largely his. I am forever grateful, David.

*Jean-Claude Kaufmann*

# Contents

# Introduction

She pulls on the stockings she's only going to wear once.

She's wearing her lacy bra, and a matching thong with flowers on it. Suspenders, heels, handbag.

Every curl carefully in place, a dab of Elizabeth Arden or Chanel behind her neck. He'll fall for her. He has to. Eye shadow. He'll look deep into her eyes and fall under her spell. She wants to be the one, the only one. The one who'll make him forget all the rest of them. And God knows he's had enough women: tall women, supple women, bi-women, older women, big-breasted women . . . Had them this way and that way, every way you can think of. She wants to be the one he can't compare with anyone else because he'll have forgotten them all.

What she really wants is to make him forget all the other women he's had a laugh with, all the other women he's danced with. Make him forget them, tonight and tomorrow.[1]

She is already quivering with emotion. The feeling of excitement is almost palpable. *Missardonic* is already on cloud nine. She talks about herself in the third person, as though she was describing a sort of different self who has escaped from her ordinary life. Tonight will be (might be) intensely pleasurable. Tonight might change her whole life.

---

[1] From the blog of *Missardonic*.

Time to go. She leaves for her date.

The clock says 19:03. He can wait a bit longer. He'll *have* to wait
if destiny is to be fulfilled. A last touch of lip gloss as she gets out
of the car at 19:17. Her lips are plumper and fleshier than they
have ever been. She walks slowly towards the bistro, letting her
heels tell everyone she is on her way, begging the men she passes
to turn and look at her. Her skirt is too short, and she can feel a
bit of a breeze. She can tell that they fancy her and that just the
sight of her long legs as she strides confidently towards another
man turns them on. They're jealous. He'll fall for her, all right.
For just one night, he'll forget everything.[2]

He's there, waiting for her in the back of the café. The photos
didn't tell the whole truth and even the webcam gave only a vague
idea of what he looked like. Now that she's face to face with him,
she gets a different impression: their bodies are talking to one
another. The waiter comes over to take their order. 'Another
mojito, please.' There's everything to play for now. She knows
that. It's not like the *commedia dell'arte* of old. The codes governing today's little plays about love are all the stricter in that the
script is open-ended. It might be just a drink and it might be just
for one night. Or it might be for the rest of her life. Who knows?
    But surely there's nothing new under the sun when it comes to
dating. It's just a first date, and there's nothing new about a first
date! That's where you are wrong. This might look like a traditional date, but times have changed and the stakes are very different these days. Only a few elements of the ritual (getting to
know one another over a drink) seem to have survived intact. But
appearances can be deceptive. The rituals remain unchanged (and
they are even more rule-governed and conformist than ever) in
order to mask the fact that things have changed. The point of all
these rituals is to ward off the fear of falling into an existential
void. It is as though we need what is in effect a new courtly code
in order to find our way in an emotional world in which no one
has a compass.
    The world of dating suddenly changed at the very beginning of
the third millennium. A combination of two very different phe-

---

[2] Ibid.

nomena (a new sexual assertiveness on the part of women, and the fact that the internet had become part of everyday life) triggered a velvet revolution. History sometimes does prove to be made up of unexpected combinations of events.

I do not have the space in this book to go into everything; the issues raised are too far-reaching. The Prologue does no more than look briefly at how people talk to teach other online. Thanks to the internet, there are now two very different stages to a date. Our lonely hearts have spent time – just how much time varies – chatting online before they meet for their first real-life date. The first part of the date is, in psychological terms, very relaxed. Anyone who goes online enjoys great freedom. They can say things that they have never dared to say before. They can cheat and, most important of all, they can break off the relationship when they see fit. They don't even need to apologize. It is the consumerist dream of modern times: take without being taken in. It is only when love comes into play – if only in a modest way – that the dream proves to be an illusion.

# Prologue: On the Net

## Love's new world

Something really did happen at the turn of the millennium. The atmosphere suddenly changed. Bestsellers like *Bridget Jones's Diary* and cult series, such as *Friends* and *Ally McBeal*, were unashamed celebrations of the single life, as were the trendy urban games based upon the 'search for a soulmate' (speed dating). At the same time, the number of computer dating sites, which first appeared in the mid-1990s in the United States (but which only developed slowly because of their technical limitations), literally exploded and increased their turnover tenfold in the space of three years (Online Publishers Association 2005). The world of dating suddenly changed. The internet bubble made us all very daring. Unfortunately, the bubble burst. The energy could not be sustained, and 9/11 revived our fears about security. The change of atmosphere had been nothing more than a cheerful parenthesis. But once that parenthesis had been closed, the internet revolution continued at the same pace as before. The real change had already taken place: the landscape of dating had changed completely. Everything was now just a click away.

In the first edition of *The Single Woman and the Fairytale Prince* (Kaufmann 2008 [1999]), I analysed women's expectations and the new difficulties they experienced when it came to commitment. The internet then caused such an upheaval that I had to update my book. In the second edition (Kaufmann 2008 [2006]), the emphasis was on the new possibilities opened up by online dating.[1] But I then had to go still further and to demonstrate that the internet revolution has also revolutionized what surfers call face-to-face real-life dates. That is the goal of the study described here. It can also be read as a sequel to *The Single Woman and the Fairytale Prince*.

At the moment, the number of dating sites is increasing rapidly and they are generating considerable profits (they have an annual growth rate of 70 per cent in the United States: Belcher 2006), even though they face growing competition from free sites and even though a new trend has emerged in sites that specialize by bringing together specific types of users (defined by race, religion, occupation, affinities and so on). They also face competition from messaging networks and blogs, which make it possible to talk openly and to become intimately involved in someone else's life. The future will therefore probably be very different from what we are seeing today. But, whatever innovations the future might bring, the important thing is that dating via the intermediary of a computer screen has become not only widespread but common-place in a very short space of time. The way in which it has become part of everyday life has been analysed by Robert Brym and Rhonda Lenton (2001), who demonstrate that the use of these sites is spreading rapidly and that news of them is passed on by word of mouth within networks of acquaintances. New users quickly overcome their inhibitions and criticisms and rapidly convince their new friends that they too should 'join the club'. Online dating, whose image was once little better than that of marriage agencies, has, in the space of only a few years, become a normal and legitimate way of finding a soulmate. And as we shall see in this book, it has become a normal and legitimate way of finding a sexual partner – long-term or otherwise. It has even become trendy, which is not something that can be said of marriage agencies. Agencies tend to appeal to a public that is rather traditional, rural and mature, and are the last resort for those who are desperate to find a partner. Those who visit dating sites, by contrast,

[1] The following pages either draw on or are inspired by the chapter on 'The Internet Revolution' in *The Single Woman and the Fairytale Prince*.

tend to be young, highly educated people who live in the cities and take part in a lot of social and leisure activities (Brym and Lenton 2001). They are open-minded about change and are, for example, more likely than most to be in favour of women's rights and sensitive to anti-homosexual discrimination (Madden and Lenhart 2006). They are by no means as lonely and desperate as we sometimes imagine them to be. The other reason why computer dating has spread so quickly is that it has been imposed from on high as a model for youth and modernity.

Its popularization has been so smooth that the internet revolution looks like no more than a peaceful and essentially technological change that is far removed from 'real life', which appears to go on as it always has done. The change is in fact much more radical than that, and the internet really has ushered in a very different age of dating. Dating is easy and intoxicating, but it is full of hidden traps that can make it even more difficult to find love.

## The hypermarket of desire

Women were a little suspicious at first, but lots of them now use the net. How can a woman resist? All it takes is one click. It takes only one click to see a succession of men, and more men – hundreds of them. They are smiling, pleasant and available. They put their masculinity on display, tensing their muscles in their swimming trunks or proudly showing off their leathers as they pose on their bikes. A click is all it takes to choose one. Welcome to the consumerist illusion which would have us believe that we can choose a man (or a woman) in the same way that we choose a yoghurt in the hypermarket. But that is not how love works. Love is not reducible to consumerism, and that is probably a good thing. The difference between a man and a yoghurt is that a woman cannot introduce a man into her life and expect everything to remain the same. A man will turn everything upside down, and she will never be the same again. And nor will he, come to that. Both their identities will undergo a metamorphosis. And that is both irresistibly attractive and terrifying.

The internet gives the opposite impression. For a man or woman who is sitting quietly at home in his or her slippers, unshaven or with no make-up on (if there is no webcam), the great advantage of making contact on the net is that it all feels so safe. She can log on with one click, and log off with another click. With one click she can look at a profile and then close the page with another click. She can send an e-mail with one click and, if the message she gets in return does not appeal, she does not even have to reply. An individual armed with a mouse imagines that she is in complete and absolute control of her social contacts. She does not realize that she is becoming caught up in something that is beyond her control and that she will not emerge unscathed. It is all very exciting to begin with. All the usual obstacles appear to have vanished, and a world of endless possibilities opens up. It is as though all she has to do is to pick and choose in a magical wonderland. A woman on the net is like a child who has been let loose in a sweetshop.

And yet the first difficulties appear very quickly. They have to do with search methods. The techniques that guide the novice through the sites are very effective. But some questions remain unanswered: how does love work? Is she in some way destined to find her soulmate? Will she know intuitively, or will the first messages they exchange tell her in a flash that she has found the man Love meant her to meet? Paradoxically, the internet reactivates the idea that 'somewhere, it is written'. Jennifer, also known as *Cinderella69* (she was born in 1969), dreams aloud in her blog: 'I'm telling you, this is Love Year Zero, the Year of True Love, the Real Thing. You couldn't do this until now. You went on waiting and waiting for your Prince, and you still had a long wait ahead of you. Because he didn't know you were waiting, poor thing! Now you've gone on the net, and everyone knows it. If it is written somewhere that you will meet him, there are no more excuses. It can't fail to work. All you have to do is look.' In the old, pre-net society of the twentieth century, 72 per cent of us met our partners at school or university, at work or in our networks of family and friends. In that sense, the internet really is revolutionary because it makes it very easy to make contact with people we do not know. For singles who live outside the big town centres or who have been marginalized for some reason or other, it represents an unhoped-for tool, and a truly magical opening onto the outside world. The net already has its stock of wonderful legends, like the tale of the disabled couple who would never have met without it.

But let us reread very carefully what *Cinderella69* told us. Although she is firmly convinced that she is destined to meet someone, she ends by saying 'All you have to do is look around you.' Fate needs a helping hand. The internet only works if we make active use of it. And it is, of course, when she begins to look around that things start to go wrong. The reversal is spectacular: the very thing that was so exciting (the vast number of men on offer) is now mentally exhausting. There are too many of them, and too much choice makes it impossible to choose. *Channelchris* feels that her head is spinning: 'In any case, when you look at their profiles, they're all the same. Charming, sporty, generous, funny, "no mind games", good looking, sensual . . . They practically guarantee that you'll be on cloud nine. Everyone's a winner. Let's give it a go!'[2]

## A virtual slap in the face

Computer dating is attractive for two reasons. The possibilities are endless (everything seems to be both possible and easy). In psychological terms, you feel perfectly safe, so long as you stay in front of your screen. The feeling of safety is in fact relative, as the internet is not as virtual as it is often said to be. Once contact is established, the relationship is real, and the distance changes only the way it works. It is, of course, easier to back down (either by making an excuse or by saying nothing), and this does make very intimate exchanges easier, but doing so is not without its repercussions for the man or woman who is on the receiving end. Many people prefer to meet on the net because they are afraid of being rejected 'in real life'. Unfortunately, they are even more likely to be rejected on the net, and the rejections can hurt. The man or woman who does not know how to break things off quickly enough is immediately trapped. *Channelchris* learned this to her cost.

> Before, I used to reply to everyone out of politeness. It seems that doesn't happen very often on dating sites. You have to learn to delete them very quickly. Now I understand why they wouldn't let

[2] Comments from her blog, *Journal de mes rencontres sur internet*. Christelle has also published a book based on her blog (Masson 2006).

me go and sang my praises. 'Thanks for replying. It's unusual for
a chick to reply. Thanks, many thanks. Someone nice at last. Let's
get to know each other better. We can't leave things at that.' And
so on and so on. 'No', I said, 'No.'[3]

Some people are inundated with offers, while others are suddenly
dumped. The internet is like everywhere else: you can be slapped
in the face.

The anonymity is relative too. *Channelchris*, who is looking for
the love of her life, reacts when she comes across her 'exes' yet
again. 'All present and correct. *Summersun. Hope62, Bond-008,
Homerus, Ace of Hearts, F-Sharp* . . . So it didn't work out the
way they wanted it to. Obviously not. Didn't work out for me
either.' Their fruitless quests are a reflection of her own failures.
The endless profiles are beginning to look less magical. A month
later . . .

Click . . . Oh no, not him again! *F-Sharp* . . . Oh no . . . Still signed
up. That's a bit much coming from someone who's always said
no to 'dating to order'. I could have done without that. That will
teach me to be curious. Mind you, it's just like passing an ex in
the street. Except that on a dating site, it's there in black and white
that he's still single. And except that he doesn't want to go on being
single.

The internet has a long memory, and every click leaves a record,
even when we try to delete it.

Breaking news: *Hope62* is single again. He's back. Not a comma
in his profile has been changed. Fancy that, I'd forgotten that he
claimed to be 'shy'. That's one way of putting it. For the benefit
of new readers, this was the famous date on 8 October. Not something you forget. So, the thing with his *Girl from Toulouse* must
have gone belly-up if he's signed up again.[4]

The traces they leave behind are especially damning for anyone
who really is looking for a soulmate, especially if the search has
been going on for a long time and seems to be getting nowhere.

---

[3] *Journal de mes rencontres.*
[4] *Journal de mes rencontres.*

For official purposes, the net is free and tolerant, but the criticisms can be violent. *Q-Tip* knows all about this.

> Being single is like being unemployed. The longer you're out of work, the more reluctant employers are to give you a job. Oh yes, being unemployed for too long really is something to be worried about. Being single is the same. The longer you stay single, the fewer guys you attract. Because there's something fishy about a man or woman who has been on their own for a long time.

Is it better for your image to describe yourself as someone who is just looking for a good time and not a soulmate? Not in the case of women, as we shall soon see.

## A new drug

Once the excitement of the first months is over, disillusionment begins to set in. The would-be dater feels she has had enough. And yet she still finds it impossible to tear herself away from the computer. The attraction is too strong. Even though it becomes nauseating, virtual reality proves to be less virtual than it seems because thousands of tiny threads irresistibly bind us to the screen and give us the feeling that we exist and are recognized as existing, and even that we can expand our horizons to infinity. How could anyone turn down such an intensification of the self? Online dating quickly becomes a drug, and we cannot do without it. Pascal Lardellier (2004) describes the stages of cyber-dependency. The worst addicts are those who are not good at relationships in the real world. The computer gradually becomes the most important thing in their lives and takes over completely. It becomes an obsession. They have to log on as soon as they get home.

Increasingly, we live in a world of addictions. This is because autonomous individuals who are doomed to construct their own lives need to be supported, to have people around them, to have cuddles and to be swept off their feet. Cocaine, alcohol, tobacco, sex, work and telly can all be misused and taken as drugs. For some people, visiting dating sites is a drug, and there is nothing soft about it. The exasperated *Channelchris*, who dreams of 'getting back to real life', 'staying in bed all Sunday morning' and 'making lots of chocolate cakes', took drastic measures. 'My computer is on holiday too. Complete rest. I've uninstalled MSN, and that's saying something . . . I've even put a sheet over it, so that it looks like a cage where the budgie has gone to sleep. That way, I'm not tempted. *Vade retro Satanas!*' Two days later, the computer was back on again.

It is also possible to become addicted to a compulsive form of online dating that never leads to anything in real life. But the worst form of addiction has to do with the initial phase of making virtual contact. As I have already said, the internet means that there are now two stages to a date. It is as though the first real-life date marked a new beginning. Because they want to see it as a seamless process, two-thirds of those who use dating sites send photos, and 86 per cent make phone contact. A few turn on their webcams (Brym and Lenton 2001). There is also a growing tendency for people to get ready for a date by investigating who the online contact really is, without saying anything to him or her. How? By going on the net, of course, and by using search engines to follow up the clues he or she has left. Anyone who uses the net leaves countless clues that make it possible to reconstruct their history and to discover various facets of their personality.

These searches become more common when someone is seriously looking for a life partner. When they are just interested in having a good time, the stakes are not so high, and the surprise effect might even be quite pleasant. But who is looking for what on the net? This is where everything becomes confused. Men, who have always been very keen on the idea of sex with no strings attached, often use false identities. And in the meantime women, who were until recently quite shy about this, are increasingly tempted by the idea of a one-night stand. Even *Channelchris*, who does not have a one-track mind and who is hoping to meet the love of her life, is sometimes tempted. She finds *Fireblade11* quite tempting: ' "I like briefs or Y-fronts but never boxers. I hate clothes that are loose-fitting. I've nothing to hide, and I don't need to cheat." He's got some nerve, this "*Fireblade11*" . . . Hmm . . . Calm down.' She was very tempted, but five days later she is not so keen on dating *Fireblade11*:

'So, you like my profile. Do you? Let's get to know each other better. If you don't like me, I've lots of friends who are single.' Oh no, he wants to tell everyone in the neighbourhood. Careful, he doesn't live that far away from me. One click, and I've got it all: e-mail, MSN address and mobile number. Slow down! I'm not sure that I want this. Puts it about a bit too much for my liking.

After a few disappointing experiences when the early promises were quickly forgotten about, *Q-Tip* is now openly on her guard: 'When *Serial Pick-Up Artist* asks me where I get my pretty smile from, or when *Ridiculous Seducer* pretends to be interested in what I do for a living, I quickly realize that all the *Man with Balls* is interested in is a shag.'

## What does 'just for sex' mean?

Internet dating has been with us for about ten years or so. For most of that time, the rules of the game seemed to be fairly clear. On the one hand, most women were looking for true love, for the man they wanted to share the rest of their lives with. On the other hand, most men made it clear that they did not want any emotional involvement, which meant that they were actually interested mainly in sex.

All this is beginning to change, and that is what this study is really about. First, because women are demanding the right to have a good time – and nothing more – in the way that men have been having a good time for hundreds of years. Second, because the dividing line between sex and sentiment is becoming increasingly ill-defined. Take the example of *Anadema*. He is certainly not under the impression that he is a predator. He has his flings and whilst they are sexual, they are also emotional and even relational (even though he never commits himself too far). He is surprised by the comments he gets on his blog:

Oh really . . . about my relationship with Lila . . . I've heard all sorts of things . . . 'Just another of his sexual exploits' . . . 'It's just physical' . . . 'All he's interested in is the sex' . . . 'He just wanted to get her on her back' . . . 'A fuck buddy' . . . 'Sex, sex and more sex' . . . Anyone who thinks there's nothing more to my relationship with Lila than sex is in for a big surprise because we do all sorts of other things: meals, drinks, go for walks, watch films, smoke rare plants, chat over a drink . . . and what's more, we respect each other. The emotional intensity varies, depending on the relationship, on the pleasure we get from being with someone else. It is up to us to decide when we want to call it love. We all have our own standards.

The new set-up means that it is no longer possible to make a clear distinction between serious dating and purely sexual encounters. Both extremes are obviously still there. But, increasingly, there is an intermediary zone in which nothing is really preordained. The Date (a magical moment that comes once in a lifetime, and that is highly idealized because the stakes are so high) has given way to the simple fact of meeting someone in a much more concrete and banal sense. No one knows in advance what is going to happen. That will be decided over a drink, and it will depend on how the people involved feel. The way we initially make contact with people on the net is, logically enough, beginning to change too. Once upon a time (only ten years ago), the marriage-bureau principle still operated: everyone was quite clear about what they expected of their ideal partner and set out their selection criteria. But now, just talking to people is what matters. We want to get to know each other, tell each other things on the net. Sometimes, we want to have fun together. And sometimes we might want to talk dirty for a moment or two. 'The Web has taken over because it is the biggest brothel that has ever existed' (Brym and Lenton 2001).

Dating sites (which are officially meant for people who are looking for the ultimate big Date) are being subverted by these new demands, and the same demands are increasingly appearing in blogs and social networking sites.

## The net's hidden treasures

Even before we become involved in anything sexual or even just amorous, the web has a lot to offer. Using the web is an experience in its own right: with just one click, we can have a life that is more real, more intense and multidimensional. When we go online, we become part of a vast network of acquaintances. We may or may not be able to identify them, and we may not even know them well. We can establish fleeting friendships and temporary alliances. We

can talk about how we feel – and a lot of people do so – and
exchange signs of affection. There is something heartening about
all these virtual kisses. There is an obvious desire for closeness and
tenderness on the net. The cruelty of online zapping contrasts with
the underlying search for humanity and generosity, and the hyper-
sensitivity. Hence the inflationist use of smiley faces; such little
symbols come in very handy when we do not know what to say to
each other. And there are a lot of would-be writers, poets and
artists of all kinds out there because the net gives them an outlet
for their talents. Astonishingly, love letters, which we thought had
gone forever, are making a comeback after 100 years of oblivion.
Just listen to *Q-Tip*: 'I spent yesterday evening with *Rainbow Scarf*:
his poems left me all of a quiver.'

Everything is possible: we can have friendly chats and we can
talk about love. We can enjoy long-distance romances or develop
a bulimic addiction to one-night stands. We can also flirt online.
It is risk free. The stakes are not as high as they are with real-life
dating. And the disappointments are not so great either. But the
content can 'become very hot when you click with someone'. One
quarter (most of them are women) of those who visit dating sites
never go out on a real date (Brym and Lenton 2001), and online
flirting is the most popular form of erotic activity. But for most
people, a real-life date is the big event. Once we are together in a
physical sense, everything changes. And if the love machine does
get under way, it is no longer so easy to stop it. Relationships that
begin like this can quickly become very intense. It might last, or
it might not. We don't know if is purely sexual, or if there is more
to it than that, and we don't know where it will lead. The problem
is that nothing is written in advance, and that the future is completely open. Fortunately, the very conventional side to the ritual
calms these fears and conceals the indecisiveness. The very banality of the codes is a form of protection: we order a drink, chat
and smile at each other.

This is in fact a social construct: the codes are banal because
they are the product of a process that made them banal. They are
meant to be banal. I thought that it would be a good idea to begin
with a very concrete description of this new, and deceptively
simple, ritual, before we look at the underlying issues. This will
give us an insider's view.

Get yourself ready. Your date is waiting.

# Part I

# In Real Life

# 1

# 'You Never Know What to Expect'

A first date is like a new beginning. All the online outpourings that came before it have been almost forgotten. This time, the practicalities are the only things to be talked about: where and when? This marks a sudden and unspoken break with the past and ensures that anything can happen. Both parties feel that everything is about to change, now that it's happening in real life.

And yet, what has been said and written cannot be deleted just like that. They have already committed themselves, and they are both involved in something that is already largely predictable. This is especially true if their online flirtation has taken on overtly sexual overtones. How can anyone believe that they are just going to chat about this and that after all those online cuddles and all that intimacy? But look at what happened to *Thirty-Something in the Jungle* when she went on a date with *Striking Seducer*.[1]

## 'I felt a bit like a call girl'

'I'm not sure that you will really understand the madness that came over me. You're going to think I'm a complete nympho.'

---

[1] She has not made up this user name. Her blog describes the 'trials and tribulations of a thirty-something-year-old in the jungle of internet dating'.

*Thirty-Something* had recently signed up with a dating site. She was disappointed. All the clichéd phrases, the awful stereotypes and the spelling mistakes were signs that the conversations were going to be both mediocre and repetitive. And then the online chat became more erotic. She hadn't gone online to look for sex. But there was something irresistibly attractive about the realization that she could attract men. 'Gradually, I realized that that's what I was looking for . . . I find it reassuring to know that men find me sexy.'

It was *Fantaisy* (sic) who introduced her to this world.[2] Directly and crudely. He was only interested in sex, but she was still tempted when she saw his photo (which was not posted on the site; most married men, and there are a lot of them on the net, do not post their photos[3]). Very good-looking. But when it came to a date, she took fright and stood him up. She had no such hesitations about *Striking Seducer*. 'I let him take charge, just to see where he would take me. And then it started! No one had ever talked to me like that. Line after line of poetic–erotic suggestions. Nothing soppy. Far from it. This stuff would shock anyone. Well, it did me, at any rate.' There was a brief interruption when she had to move house. After that, her online love life was better than ever.

[2] Everyone chooses their own user name, and the choice of name often tells us a lot about the identity they dream of having. The best blogs also have a narrative logic of their own. Day by day, they tell true stories taken from real life, complete with characters: the heroes of the dates. Significantly, these characters are not referred to by their user names; the narrator gives them new names, as we shall see when we look at the adventures of *Herring* or *Soft Toffee*. The humorous connotations of these colourful names often mask an underlying violence. This is because negative remarks can do the rounds, both on the net and in real life (the net is not as anonymous as it might seem) and can cause a lot of damage if they pinpoint real faults. In the case of *Herring*, that did not matter too much. He was just passing through, and acquired the name simply because he was Scandinavian (he did in fact complain to me, and asked to be renamed *Shark*). In the case of *Soft Toffee*, it did matter: he was publicly humiliated because his sexual performance was so poor.

[3] Some 20 per cent of the men who visit dating sites are married (Madden and Lenhart 2006), and they are looking for sex, and not a soulmate. Posting their photos would obviously give the game away (it is only because there are strict controls on anonymity that the internet remains anonymous). They will send photos on request, usually by e-mail.

Hours of chat over the next three days. Fantasies I'd never dreamed of. He was driving me wild! One night, he asked me to call him and I gave him my mobile number. He was as good on the phone as he had been in the chat room. An incredibly sensual voice, and he called me 'darling'. I really fell for him. If he was as good at making love as he was at talking about it . . .

The idea of a date came up quite naturally, and they both knew perfectly well what was going to happen next. They arranged to meet at a hotel. 'And so he asks me to go to Paris! I wasn't too keen on that, and said I'd rather he came to Marseille. It's true, I feel more confident when I'm on my home ground.

"No problem."
"When?"
"This weekend, obviously."

It took him less than five minutes to find the address of the Sofitel.' *Thirty-Something* had been swept off her feet by a whirlwind. It was as though she was in a movie.

So off I went . . . I turned up at the hotel reception. You get the picture: Sofitel, a classy lobby, all very quiet, heels, a little black skirt, a sexy top (but not too sexy), with a black mac to cover everything up . . . I felt a bit like a call girl . . . it felt funny, spending the weekend a short walk from where I live, but to hell with the conventions.

As we shall see, what happened next was not quite what she had expected.

## 'I'm looking for a man'

In the meantime, *Marion* will help us understand that sex is not the only thing that can determine in advance the way a first date will turn out. Stories like hers are not common, but they are interesting because they show that the net can be used in many different ways, and that they have little to do with the established dating code.

*Marion* had launched an appeal on Twitter: 'I'm looking for a man.' Nothing unusual about that. She was inundated with replies from men offering to give her a good time. Nothing unusual about that either. But there was something unusual about one reply: *A Man* set up a blog especially for her (*A Man for Marion*). Their relationship, which had yet to get off the ground, immediately became public knowledge. Because everything was out in the open, there were three sets of people involved: our digital love birds, each with their own style; their friends, who played an active role in these preliminary manoeuvres (*Marion* put 'detectives' on the trail of the real-life *A Man*, and told him online what they had learned about him); and everyone else who read both blogs, offered advice and encouragement, and added their comments – both critical and encouraging. A lot of the details of their date were discussed in public. *Marion*, for instance, raised the issue of what she should wear. *A Man* didn't show much interest, but the same could not be said of her girlfriends. 'That was the crucial question on Twitter just then. A skirt? A dress? Whatever else you do, nothing too revealing, *Marion*.' The date very quickly became a major event, and that increased the emotional pressure. Carried away by this irresistible polyphonic narrative, *Marion* was becoming lyrically optimistic.

> The plot thickens. Will it have a happy ending, with a kiss on the Pont des Arts, like in the movies?
> *A Man*, *A Man*, *A Man*. I can't wait for Friday, and at the same time I'm worried about what might happen.
> *A Man*, *A Man*, *A Man* . . . will you be my man?
> I hope you're affectionate as well as romantic, because I'm a Free Hugs girl, can't help myself.[4]

They quickly arranged a date for a Friday night. *A Man* proved to be very organized. He was poetic but punctilious, and took care of every detail, even though he denied doing so. He checked

[4] She is obviously referring to the Free Hugs Campaign; members offer to hug strangers in public places. They make it clear who they are by holding up signs.

the weather forecast: it was going to rain. So they had to find a suitable place to meet. Somewhere under cover, but somewhere where he could smoke, as that was one of his bad habits. His other failing was that he had no enthusiasm for never-ending discussions about hairstyles and what to wear. After a long discussion on her blog, *Marion* had settled for a Liberty skirt. That is how *AMan* found out that a Liberty skirt is a floral print. 'Tomorrow night, we can drink cocktails and nibble a few tapas if we feel like it. The weather seems to be improving, so that gives me the chance to undo a shirt button, and you can show off your skirt. If you're wearing a floral print, I don't need to bring you flowers. It would be too much of a cliché anyway.'

The Friday they had been waiting for finally came. *Marion* was, as promised, wearing her Liberty skirt. 'I knew you'd be wearing that. I sort of knew you had green eyes, but I'd never really imagined . . .' They headed for a tapas bar. 'We got to the bar, and ordered margaritas and a plate of tapas. The tapas came before the drinks. We dug into the tapas and almost choked because the mushrooms were too spicy.' First topic of conversation: their life stories, as told in their blogs. 'We talked a bit about our little virtual games; we'd both been overtaken by events, but we'd had a good laugh writing all that stuff.' They joked, decided that the mojitos were definitely too strong ('So we'll have margaritas'[5]), and chatted about the Liberty skirt, about the art of picking men and women up in bars. 'I told him that I'm a romantic, but he didn't believe a word of it. Then we made a pathetic attempt to ask the barman for a drink in Spanish: he didn't understand a word we said. The margaritas were beginning to kick in, and we moved on to more personal topics.' Their online history meant that they had a lot to talk about, and a lot to laugh about. They were so caught up with each other that they were able to ignore the fact that there was little content in what they were saying. They may not have been forgotten about, but their virtual commitments were beginning to look less definite.

What happened next was easier. As we shall see, there has been a lot of discussion about whether a couple should kiss on their first date. *AMan* and *Marion* had no doubts about what to do

---

[5] This is obviously a personal opinion. Margaritas are not necessarily less alcoholic than mojitos.

next. 'According to the police, I gave him the look that means the
girl wants the boy to kiss her. According to the organizers (me),
he just asked me: "What shall we do now?" I didn't say anything.
We kissed.' *A Man* confirms her version of events: 'Then there was
a moment's silence. You looked at me with your big green eyes,
and then it was our first kiss.'

If they do kiss, the next big question is whether a girl should
go to bed with the man on their first date. But there were limits
that *Marion* would not go beyond. 'Just in case anyone's interested, I'm an old-fashioned girl. Never on a first date.' *A Man* may
have been a little disappointed.

'You're right, *Marion*, I'm beginning to think you are a
romantic.'

'Oh, I've got a sweet tooth, that's all' (points at her *mojito*).[6]

Presumably this merely delayed the inevitable (until the next day).
At this point, I can, unfortunately, only speculate. 'Romance 2.0'
was suddenly shrouded in secrecy. 'What happened next is between
the two of us and can't be written down, or at least not
here.'[7] *A Man* stopped blogging, and *Marion* never mentioned him
again.

We will never know what happened next. But everything suggests that Marion moved on quickly, as *A Man* was not her only
hero. Over the next few days, she dated *Stupid Prick I*, and then
*Mr Rummy* (the one who invites girls back to his place for a
rum . . . and maybe more). *A Man* was already ancient history.
*Marion* obviously has a very strange idea about what 'romantic'
means; her romances do not last for long when new physical
pleasures are on offer.

Even when the date is organized in advance and when the preliminary chat limits the possibilities, nothing is certain. In the case
of *A Man* and *Marion*, their fine 'Romance 2.0' came to an abrupt
end. In the case of *Thirty-Something in the Jungle*, what had been
said on the net held out the promise of a night of passion. Not
the shadow of a doubt about that. When we last saw her, she was
in the Sofitel's lobby, and all turned on.

---

[6] Feeling the need for some Dutch courage, they had now moved on to *mojitos*,
despite their earlier reservations.
[7] Taken from the blog, *A Guy for Marion: the Story of a Romance 2.0.*

I asked for the gentleman's room number.

'No, I'm sorry madam, there's no one of that name staying here. Have you got the name right?'

Gulp.

'Is there any message for me?'

(The receptionist looks for a message).

'No, no message, madam.' (If she doesn't stop calling me 'madam' in that sympathetic tone, I'll ram her pen down her throat.) 'Are you sure you've got the right Sofitel? . . . There's the Palm Beach on the Corniche.' *Thirty-Something* texts *Striking Seducer*. There must have been a misunderstanding, she thinks. She gets the answering service: 'Voice box full.'

'Second text message: 'Is this your idea of a joke?' I go for a coffee in the hotel bar, very much on my dignity. Still no answer. Third text: 'Prick!' I pick up my bag and walk out of the hotel, mustering as much dignity as I can. I might be wearing heels, but I'm home in a flash. I'm livid with anger. How would I let myself be taken in like that? Really, how could I believe that a guy is going to come down from Paris and ask me to spend a dirty weekend in a Sofitel hotel, just because all that sexy chat and flirty talk got me excited? As *Maria* puts it 'Nothing sexy about guys who stand you up.'[8]

As soon as she got home, she went online and found to her amazement that *Striking Seducer* was also online, and chatting to other virtual victims.

I took a deep breath, and decided to send him a killer e-mail. That made me feel better. This is just a sample: 'Is that what turns you on? Making a girl feel horny and then leaving her looking like a fool while she waits for you in a hotel lobby? What's the point of all the smooth talk, if you're not even going to reap what you sow. It doesn't take much to satisfy you. A virtual fantasy . . . that's what you want. Sitting in front of your PC and chasing women on a screen. Convincing a girl you're a witty poet when you talk to her on the phone. What a saddo. You're pathetic, despicable.'

Even when you think it has been settled in advance, and even when it is highly structured by reassuring rituals (such as meeting

---

[8] Comment on *Nire Grinak*'s blog.

for a drink), there is no telling how a date will end. A date is a specific and open-ended event, and it marks a departure from earlier online commitments. Remember *Fantaisy*. *Thirty-Something* did not dare describe her sexual suggestions because they were so dirty. He asked her out on a date ('Just a drink') and promised that: 'I won't jump on you.' *Thirty-Something* agreed to meet him, and then changed her mind at the last moment.

No matter what has been written online, and no matter how daring the phone conversations have been, going out on a first date is always a completely new adventure.

# 2

# First Steps

---

## The sound of heels

First, she has to get ready. Her clothes have to fit in with what she expects to happen. When she still believed that she was going to have a good time with *Striking Seducer*, *Thirty-Something* thought of everything: 'Things to do before the weekend of a lifetime: buy some ultra-sexy underwear, beautician, hairdresser, little black skirt, the lot! Even suspenders . . . you get the idea. And then I remembered that I didn't have any; I really must get myself properly kitted out if I'm going to make a habit of this! It will always come in handy!' And then, when it's time, get to wherever it's all going to happen.

No shaking with nerves. Wear a big coat or shoes with heels that make a lot of noise if you need to boost your confidence. *Fanette* clings to these little points of reference to ward off her feeling of panic.

So I was going to walk confidently when I got there, self-confident and walking tall. That's what I thought! And to make things even worse, I was wearing flat shoes that day. When I'm wearing heels, you can hear me coming and that gives me confidence. Stupid but it feels good. I'm wearing a long coat (it makes me look a bit dumpy

but I feel good in it). I go through all the things I've got going for
me in my head, and I say to myself that I'll do all I can to make
myself look bright-eyed and bushy-tailed.[1]

In many cases, the initial contact is not always easy. They have
to recognize each other, find something to say without stumbling
over their words, and avoid all those dreadfully conventional
platitudes. *Nathalie* can hardly manage it. 'We'd arranged to meet
outside a restaurant. I was terrified, and had absolutely no idea
what he looked like. Then he turned up and shyly asked if it was
me he was looking for.'[2] They were meeting for a meal rather than
just a drink, and that usually involves much more commitment.
Sadly, the initial feeling of unfamiliarity defined their whole relationship and they never got rid of this chilly feeling. 'We had lunch
together, and I found the idea of having a meal with a complete
stranger absolutely ridiculous. All the fine discussions we'd had
on the net, all the laughs . . . all that seemed to be a thing of the
past.' And that is as far as it went.

But first of all, it has to get off the ground. And it is not uncom-
mon for people to have doubts. There seems to be no one here. Is
this the right place? Have I been stood up? In many cases, this
simply means the two people involved are out of synch: one person
is on time, and the other is cool about these things. On the other
hand, it may be a carefully premeditated tactic. Anyone who has
been left waiting soon begins to think the worst. Waiting for a
while is fine, but how long do you go on waiting? This can be a
real problem when you have arranged to meet outdoors, or if it
is raining. In the meantime, *Flash Gordon* was nice and warm,
sitting at a table in the café. He seemed to have been waiting for
ever. 'Convinced that no woman could resist his seductive charms,'
he waited for longer than anyone could reasonably be expected
to wait: over two hours. 'I think I was even beginning to get on
the waiter's nerves by the end.'[3]

There is always the phone of course. People exchange a huge
number of text messages as the time gets nearer and as they try
to get to know each other. 'Stuck in a traffic jam. Give me five

[1] From *Fanette*'s blog.
[2] From a chat room.
[3] From a chat room.

minutes.' 'Where did we say we'd meet again?' No one thinks it is impolite to check on these things. Text messages put people at ease. Especially when they are about purely practical matters. They in fact play a much more important role than one might think. When we concentrate on the practicalities, we can forget about everything that was said on the net, and settle into the neutrality of a ritual that marks the beginning of a new adventure. We can chat and we can enjoy being with someone, but we must not give any sign that we are looking for commitment. Those are the rules. Because everything will be decided later. Over a drink. In real life.

All too often, the phone call goes unanswered. Rising levels of anxiety, disappointment and then anger. *Flash Gordon* had to face facts: she'd stood him up.

## Why are so many people stood up?

As the minutes tick by, he begins to worry: has she stood me up? And if she has, why? The longer he waits, the more the doubts undermine his self-confidence. *Christophe* had arranged to meet his date in a restaurant. That was a bad move: the restaurant was closed that day. He stood outside for what felt like an eternity. He did not have her mobile number so he could not call her. He nervously checked his own phone, but it refused to ring. He still has a bitter taste in his mouth when he remembers how lonely he felt. Why did she do that to me? Why?

It is not unusual to be stood up on a first date, and almost everyone who is stood up reacts in the same way. They want to know why. They want to find an explanation, one way or another. This is mainly because they are looking for reassurance. If they can't find the explanation they want by phoning their date, they rush home, log on, and demand an answer. Sometimes they do get an answer, and sometimes it does help them to calm down. *Antoine* found out that his date did turn up for their date. 'She got there early and saw me waiting. All at once she froze. She couldn't go through with it, so she went away before I saw her. It was the first time she'd arranged a date on the net, and she panicked. That gave me a fright. I thought for a moment that I must be repulsively ugly.' In most cases, the excuses are pathetic and it is impossible to take them seriously. *Christophe* would have been happier if she had said nothing. 'She claimed that her father was ill and that she'd had to stay with him.' So why didn't she call him to say so? 'That's the reason why I gave her my number.' She mumbled something incomprehensible. *Frogita*[4] was in the same mood. She waited in vain in the park, and it was a wet and windy day. She phoned again and again. No answer. She went home, and sent e-mail after e-mail. Still no answer. She finally got a message the next day: 'I didn't wake up in time.' They had arranged to meet at two in the afternoon.

A lot of people fail to show up for the dates they have arranged. The main reason why they do so is that, despite what everyone may think, a date is not just a follow up to their online conversations. A date brings together two people who really are different from who they were on the net. They are not more 'real' or more 'authentic', but they are different. In a sense, a date is a fresh start because the cards have been reshuffled. Which explains why the rituals have to be codified: the transition from net to real life has to be as normal and as neutral as possible.

People who try to find dates on the net can be divided into three categories (Lejealle 2008). Some chat online because they enjoy making contact with other people that way, and then decide to meet them 'in real life' for a drink . . . and possibly more. It is only when they actually meet for a drink they will find out if they have anything in common and if they are looking for the same thing. They will see each other in a new light, will have to come to terms with the fact that this is not the man or woman they were chatting with online. Those who come into the other categories solve this problem by putting all their energies into either the date or the online chat. They may concentrate on the date itself, and they usually define it in purely sexual terms. Even though they sometimes enjoy chatting online,[5] these predators live in the future tense. Inside their heads, they are already enjoying the face-to-face

---

[4] Her user name was chosen to fit in with the title of her blog: *The Toad Pond*.
[5] They are usually thinking about something else. 'I soon get tired of long discussions in chat rooms. And you have to be in good shape to do that.' From *Nire Grinak*'s blog.

encounter (or something more intimate). They make lots of prom-ises and are quite happy to lie to get their own way. Others are turned on by the magic of long-distance love.

I mention magic because modern-day love makes us feel very adventurous precisely because we are in no danger. And especially because there is no danger of commitment. This group also includes poets, people with a lot of affection to offer, and a lot of people with broken hearts. Virtual cuddles can cure broken hearts. It should, however, be noted that the fundamentalists who refuse to have anything to do with real life and who live in a purely digital world are in the minority, for the very good reason that chatting to someone online is a real experience and can be the beginning of a real relationship. The emotions we feel are very real and can, in their own way, be very intense. But those emo-tions fade when the surfer realizes that there is never going to be any follow-up, that it is all just words on the screen, that it is all virtual, imaginary and not real. He or she has to be convinced that there are no taboos, that things can always go further. He or she has to be bold enough to believe that they might make physical contact tomorrow. That in itself is enough to turn them on. They are not consciously lying. They are not lying at all; they are trying to have an emotional relationship at one remove and are therefore quite prepared to distort the truth if need be.

## Inner beauty

There comes a point when this imaginary world collapses in ruins. What happens next has to happen in real life, and she cannot make the transition from one to the other (usually because her imaginary world is so big and so solidly built). Her imaginary world may collapse when she hears his voice, or when she summons up the courage to make that initial first phone call. She stumbles over her words, which sound dreadfully ordinary. There is none of the fluidity, the humour, the poetry or the boldness she expe-rienced when she was online. A whole romantic world is collaps-ing. Even the sensuality and the eroticism have gone. It is a strange paradox: as the flesh-and-blood man or woman comes closer, our bodies go stiff and we back off.

Voices are, in any case, only of secondary importance. Images are the real, absolute enemy. It is as easy to fall in love online (you can say 'yes' with just one click) as it is 'in real life', and we usually fall in love with an image. This has to do with the new ways in which we identify the people we meet (Kaufmann 2004): as the reality of individuals becomes more complex and more fluid, the more our idea of who they are is simply and immediately determined by the way they look. That is why on-the-spot judgements are so cruel, and why they reinforce our standardized and reductive canons of beauty. Two seconds, and the die is cast. Woe betide the man or woman who does not look like the stereotypes.

I was just saying that 'real life' is not necessarily any more real than online chat. Let me give an example. Nothing could be more deceptive than a fleeting impression that is in fact triggered by social stereotypes that normalize our criteria for beauty. Intuition may well be a very effective decision-making technique, but falling in love actually takes much longer than a few seconds, as it means that we have to get to know the other person in a different sense. The point about online chat is that it allows us to escape this reductive tyranny: there are no a priorisms and we get to know the other person by discovering just how rich his or her world is. Surfers experience the joy of being free when they sing the praises of 'inner beauty'. 'Inner beauty is what matters' is one of the most common slogans to be found on dating sites. Unfortunately, the same surfers very quickly ask for photographs. They can't resist: they want to see. And no matter how strong their convictions about inner beauty, and no matter how much passion went into the online chat, a photograph immediately makes us see other people differently. No inner beauty can stand up to a photograph for very long. '*Poupidou34* told me one night, "I'm no playboy, but it seems that it is what is inside that matters. That suits me." I'm quite prepared to believe him. I guessed as much when I saw his photo.'[6]

*Nire Grinak* was let down by his photo too.

I'd arranged to meet M. even though my photo hadn't been validated. Fortunately, I checked my e-mail just before what sounded

[6] Taken from *Christelle*'s blog.

as though it was going to be a very pleasant date. 'Hello, I'm sorry but I'm going to call off tonight's date. I'm sorry, but I try to be honest. I saw your photo on Meetic. You're just not my type. I might be being harsh but I'd rather be honest. I've set up this e-mail address just to let you know. Don't try to contact me again. It wouldn't do you any good. Good luck in future. I really enjoyed the chat we had thanks to Meetic the other night. Best wishes, M.'

Being turned down like that may have hurt, but it at least (just) spared *Nire Grinak* the embarrassment of being stood up. Unfortunately, not everyone is so lucky.

Photos are not the worst thing in the world. In many cases, the photo is not a problem. The recipient somehow integrates it into his wonderful, magical world and goes ahead regardless and may even dare to arrange a date (or, which is more likely, agree to go out with someone who has asked him for a date). But come D-day, he begins to get cold feet. He is frightened, frightened of everything: frightened of himself and of his own incompetence, frightened of the coming ordeal, frightened of the other person. What he thought was going to be a continuation is the start of something very different, and he is not ready for that. He backs down. He feels ashamed of himself, and often does not have the courage to let his date know in advance. That is probably what happened to *Striking Seducer.*

Sometimes he turns up anyway. He hangs around discreetly, and watches without being seen. Once again, the image (a brief glimpse of the woman who is waiting for him) comes into play and it can be enough to kill what little passion was left. *Caro* is very frightened of the impression her weight will make. She doesn't weigh much on her blog, but she puts on the kilos when she goes on a first date. 'The only thing that stresses me out these days is me. I'm not a little slip of a thing who weighs 117 pounds. I know that some men like them bigger than that, but I'm always scared when someone looks at me for the first time.' This is not the man or woman you fell in love with on the net. He or she is completely different, a complete stranger. And this is not about a few kilos or centimetres here or there. This really is someone different, and you have to start all over again. This is someone you cannot approach.

That is why so many people are stood up.

## Dates that do not work out

Once the waiting begins (and it often does), anyone who is on a date begins to worry that she might have been stood up. Even if she hasn't, the waiting and everything else she has had to put up with mean that she is not in the best of moods.

   *Soann* stood there long enough to attract some unwanted attention.

> I waited 20 minutes for him, and that was long enough for me to be wolf-whistled, to be approached by a (really) weird guy, and to have to put up with two pick-up attempts (by two lads who saw me with the weird guy and came to my rescue), and so on. The wait was endless, and not very nice. And finally, his lordship turns up.[7]

She is impatient and thinks she is out of the woods, now that she has glimpsed the man she's been wanting to see for so long. But her troubles are not over by any means. Turning up for a date does not necessarily mean that you are really with the other person. A real date means being available, being prepared to be surprised, being responsive to someone else. Making the transition from the identities involved in online chat to real-life identities encourages us to cling to our old points of reference, to retreat into our shell and therefore to see the person we have just met in purely negative terms.

> I remember waiting for him outside my place. I was furious. Partly because I'd just discovered to my horror that the boots that had cost me a fortune didn't suit me. Partly because he was late. Partly because he hadn't made the same effort to look nice. And I thought he was too short (inevitably, given that I was wearing six-inch heels). I felt like going home there and then. I was not in the mood, and it was obvious.[8]

*Saskia* put together contradictory arguments (the fact that he was late had nothing to do with her boots) in an attempt to convince

---

[7] Taken from her blog.
[8] Taken from her blog.

herself that the date was a mistake, and she refused to accept the challenge to her identity that might make her change. She found more and more things to complain about (even the café looked dreary), and she eventually came to feel just as negative about the man she was with.

> I was beginning to find this date oppressive. I wasn't really ready for it. The conversation was not exactly lively, I did not really feel comfortable, and the café was dreary . . . empty. I asked him if he wanted to come back for a drink, but I knew that I was not in the mood for fun and games, and warned him that nothing was going to happen. When we went back to my place, I didn't want him anywhere near me. I did not feel comfortable, and the silences were a bit embarrassing.

*Saskia* put some music on to fill the silence. He did not like her music, and said so quite openly. The date was over even before it started. How could they have been so close on the net? They'd even spent a lot of time talking about music. She remembered one very suggestive thing they'd said: 'If two people agree about music, it means they can agree about everything.' In real life, they could not agree about anything. And they could not agree about anything because *Saskia* had decided from the start that was the way it was going to be.

Perhaps things began to go wrong when she got annoyed about her boots. Who knows? The fact that he was late cannot have helped. Nor did the fact that he was a few inches shorter than she had expected. When people explain why their dates did not work out, they often claim that the person they have met lied about his or her height or weight, or that they did not look like their photo. They claim to have been the victim of misleading advertising, so to speak. *Meeticgirl* carefully lists all her date's faults:

> 20.05. First disappointment. His photo must be five years out of date; he looks so old for a 28-year-old!
> 20.10. Second disappointment: he smokes like a chimney. OK, so I smoke too, but he tells me he smokes over twenty a day. Yuk!
> 20.15. Third disappointment: he gets out of the car. Oh no! The site said you were taller than that: your eyes are at the same height as mine.[9]

[9] Taken from her blog.

We cannot take this at face value. This is a facile argument. In
fact it is far too facile. Studies (Brym and Lenton 2001) show that,
whilst people often tell white lies or are economical with the truth
on the net (all's fair in love and war, and no one is going to post
their worst photograph), it is fairly unusual for anyone to be really
untruthful (and, of those who are, most are unscrupulous sexual
predators). Anyone who is looking for a proper date is caught up
in a logic of self-expression, and that encourages a minimal degree
of honesty. They do tinker with the facts, but only to a limited
extent. They know that if they cheat too much, they will pay for
it sooner or later.

The 'misleading advertising' argument is therefore grossly over-
stated. It is often a way of masking the fact that we are the source
of the problem. A real date is a challenge to our sense of who we
are. It gives us an existential jolt. This makes us feel uncomfortable if what the other person wants is not enough to convince us
that we have to change. And because we are lacking in confidence,
we always fall back on the same solution: we cling to our old
points of reference and find excuses to pull down the shutters. The
individual retreats into his or her shell, and sees the other through
a very critical prism. Two people can grow apart when they are
close together in physical terms, and the atmosphere becomes cold
and sad. 'I asked myself what I was doing there with this man I
didn't know. I wanted to be somewhere else.'[10]

So we become embarrassed. And the more embarrassed we
become, the more oppressive the silence and the more palpable
the unease. If he is more embarrassed than she is, he tries to fill
the silence by talking about anything and everything. He thinks
that nothing could be worse than this silence, but he is wrong.
The endless chatter just gives her more reason to dislike him. It
might be the way he says things. 'He was really doing my head
in with his "and then he goes . . .", "so I'm like . . ." and "know
what I mean". . .'[11] Or it might be what he says:

> This morning, he began by telling me that some President or other
> said something anti-Semitic. And then he started explaining the
> Iran-Iraq war to me . . . Even though I'd told him I wasn't inter-

[10] *Anne* (chat room).
[11] From *SoAnn*'s blog.

ested in wars and that kind of thing . . . if I sent him one text to that effect, he must have sent a dozen. And as it was, I interrupted his historical monologue. At that point, I felt like getting my things together and leaving.[12]

She ends her blog thus: 'My last-but-one date was a pain because the boy had no life, no personality and nothing to say. . . . I almost died of boredom . . . I couldn't imagine anyone being worse than that. But the next one was worse than that! Fucking hell! I couldn't believe it!'

It could have been worse. *Caro* might have gone out with someone who had nothing better to talk about than what he used to get up to with his ex. *Singlegent* still doesn't understand why some women do this.

> She told me all about all the affairs she'd had, all the dates she'd arranged on the net, all the flings she had as a result. She made me sick. And yet she wasn't that bad-looking. When we first met, I found her quite attractive in the physical sense, but by the end of the date, she was making me sick and I couldn't see her the same way. Even though she was beautiful in the physical sense, all I could see was this weirdo who was describing her private life to a stranger. You know what it's like when some old drunk tells you everything that ever happened to him as you sit in the back of the bus. (Chat room)

He paid the bill and fled. *Alexander* only had to put up with stories about one ex, but he too was just as surprised and sickened:

> We'd arranged to meet in the shopping centre car park, and we walked over to the bar-restaurant to have a drink. She immediately began to talk about her ex: 'Oh, I used to come shopping here with my ex.' And as we went into the shopping centre, she said: 'My ex used to like buying clothes from that place.' And: 'I once had a meal here with my ex.' It was the same when we got to the bar: 'My ex this, my ex that . . .' At one point I said to myself: this isn't real . . . she must be taking the piss. But there was something manic about the way she behaved, something compulsive that made me think that she wasn't doing it deliberately. Talking a bit

---

[12] *Caro*'s blog.

about an ex is one thing, but there comes a point when it gets
unbearable. (Discussion forum)

*Alexander* came to the conclusion that she was mad.

That is, of course, too summary a judgement. Feeling uncomfortable can have all sorts of side effects. And the embarrassment
that causes the verbal diarrhoea often serves to mask a tendency
to be critical and intolerant. Blaming the other person for everything is just too easy. It takes two to ruin a date. And things begin
to go wrong immediately if they are cold and distant, and if the
way they talk to each other (cold and distant) gets it off to a start
that will probably lead nowhere.

## Making a quick getaway

When that happens, all we want to do is get away. Unfortunately,
polite convention makes that impossible. When we are this close
to someone, we feel trapped. And the more we want to get away,
the more our mind wanders. As we pay less attention to what is
going on, the date is doomed even before it has got under way.
We need to get away as soon as possible. But how?

The problem becomes even more delicate when our partner
seems to be in a good mood. Running away would be more than
impolite. The woman who is sitting there is committed body and
soul, and she is doing her best to please: her smile is charming
and her eyes are sparkling with desire. The man who wants to
run away is well aware that he is going to hurt her very badly in
psychological terms. Her hopes will be dashed, and it will be a
severe blow to her self-esteem. She will feel rejected. He is going
to turn down everything that is good about her, and all the love
she was prepared to give. Suddenly, and without any explanation.
It's almost as bad as being stood up.

*Rosko* has developed a technique that may well be vulgar but
that does not do too much damage. 'The ladies might find it a bit
mean and uncouth, but I've come up with an effective way of
dealing with internet dates that are going nowhere. It's simple. If
I don't like her, I act like a jerk and come out with some crude
jokes.' He even thought of belching loudly after taking a good
swig of beer. That obviously takes some nerve. For ordinary

mortals, the usual method is well described by one of the many guides to dating etiquette to be found on the net.

1 Meet for a coffee or for lunch if you feel uncomfortable about going on a date. Choose an informal setting such as a café, and do not go out for an evening meal that might last for hours.
2 If your date is rude or unpleasant, leave immediately. You don't owe him anything.
3 If you have nothing in common, stick to your agreement and bring it to an end as quickly as you can without being rude.[13]

Stick to your agreement and keep things strictly neutral. Meeting for a drink means precisely what it says. That is what you said you were going to do, even though you might have had something else in mind. There is nothing impolite about sticking to a contract. Meet for a drink. Nothing was said about what would happen next, so you are quite free to leave when you feel you have to.

I have already described how the fact that the ritual ('just a drink') is at once so banal, so neutral and so central facilitates the transition from online dating to the new phase that brings together two secret identities. We now find that the ritual also offers some protection to individuals who want to escape the hellish spiral of an unwanted commitment. Meeting for a drink is not a meaningless ritual. More to the point, its very banality is an indication of its functional importance.

## Having a drink

Having a drink in order to set the seal on a friendship is an age-old ritual. In olden days, having a drink together (or sharing food) allowed many tribes to avoid war and to forge alliances. The ritual takes different forms in different times and cultures, and those forms are highly coded. Going for a drink on a first date is

---

[13] Taken from la rencontre.info.

an extension of this tradition. It has, however, taken on completely new meanings, and the codes have had to change accordingly.

Without wishing to undertake an encyclopaedic study of the subject, I do think that we need to establish that meeting for a drink can have two different meanings at two different points in time. The date in the café may not be the end of the evening, even though nothing has really been decided yet. If she lets him walk her home and then asks 'Would you like to come in for a coffee?', she is saying that she will put up no resistance: this is an open invitation. The very banality of the ritual makes the transition easier. But in this case, it signals that a decision has been made, and not that everything is undecided. In both cases, it is something as pathetic as a drink that allows us to take the next step, as though we could forget both what is at stake and all the hesitations and fears. *Pinklady* even describes a date with an advertising executive. He was drop-dead gorgeous, but he had surprisingly little to say (perhaps because he had nothing to sell). Two drinks were all it took. He did not say much and he did not waste much time.

> We had a glass of champagne in a bar and then he asked me back for a drink. As a rule, a woman who agrees to go back with someone for one last drink knows what she is letting herself in for, always assuming that she's not a complete innocent. So, we had our night of passion but that's as far as it went. (from her blog)

*Frogita* gives a rather unusual counter-example. Someone was pestering her with text messages, asking if they could 'have a drink together'. She said that it really would be just one drink and nothing more. She got an immediate answer: 'OK, no problem, some other time, then.' 'So I said to myself, some guys really do know how to make a girl feel special.' Her pest was confused, and had failed to realize that a first drink and a second are two very different things. A first drink is like a transitional stage; because it is neutral, it means we can move on to the next stage, but nothing has been decided yet. Because we are concentrating on the details of the ritual, we have time to decide if are going to take the plunge or not. The ritual itself has to be banal, and we make an effort to keep it that way. Both parties are so intuitively aware of this that they sometimes choose to have a very ordinary drink in a setting that is just as ordinary. *Anadema* remembers a date with *Celia*.

> We walked up and down the boulevard, looking for somewhere to have a drink. She suggested going to the Paradis du Fruit, but it gets very crowded and it's not a very good place to try to get to know someone. I suggested the café opposite: it was as grubby as you would expect a bar in Paris to be, and the obvious absence of customers guaranteed we'd have some peace, even though the standard of cleanliness might leave something to be desired.

What happened next more than confirmed his first impression.

> The landlord, an old guy who was in a spectacularly bad mood and rude into the bargain, brought us our drinks. I had a coffee and *Celia* had what he claimed was a cocktail. He slammed it down on the table and spilled some of it. A drop of vodka, orange juice made from concentrate, served in an ordinary glass. What do you expect for ten euros these days?[14]

Fortunately, that gave them something to talk about. When they later exchanged text messages, the horrible cocktail actually served a purpose by giving them something to laugh about.

But at the time, the off-putting look of the place could have had a disastrous effect. That is why *Maylis* prefers to choose places that have a certain charm. If you go out on a date, you expect to have a nice time. So why be content with an ordinary café (to say nothing of a sordid one)? And if your date finds the enchanted spot, so much the better! That is what happened with *Bitter Toffee*. 'He arranged to meet me at the Tea 4 Two, near Bercy Village. I'd never been there before, so it was a nice surprise. Comfortable, cosy atmosphere, electro-pop playing softly in the background, and a good selection of teas.' *Maylis* was won over immediately, and the atmosphere probably helps to explain why she couldn't take her eyes off *Bitter Toffee*. 'He looked really cool, and was wearing a coloured scarf to die for. He had a suave sensual voice, and he knew how to use it to charm the pants off a girl. Lots of little smiles and suggestive winks. Yes, I have to

---

[14] From *Anadema*'s blog.

admit to having something of a weakness for smooth talkers. That's why I get disappointed so often.'

*Bitter Toffee* was yet another disappointment. Which makes one wonder if a nice place really is the best choice. Perhaps the place makes the decision for us, precisely because it is so nice. Those who argue the case for meeting in an ordinary café might be right after all.

## Who pays the bill?

Because the situation is so banal (they obviously quite fail to see that it is a social construct), lovebirds try to get close to one another as they talk, but they are also quietly wondering about what is going to happen next. And it does not take much to disrupt the 'first drink' ritual when the time comes to leave the bar. Every detail of the ritual seems to be coded but, strangely enough, one point remains ill-defined: who pays the bill? If they were just meeting for a drink, the question very rarely comes up when the date is arranged (if it involves going to a restaurant, a formal invitation is more normal). Chivalry suggests that the man should pay. But after several decades of feminism, that might look like an affront to the woman's independence, and might have unacceptably patriarchal overtones. Skinflints do play the equality card, but women are less keen to play when the time comes to settle the bill. So the equality principle suggests that couples should go Dutch. That is not always straightforward on a first date, as there are two ways of dividing a bill: either both partners pay for their own drinks, or they divide the whole bill by two. In a fit of generosity, and breaking the habit of a lifetime, *Anadema* opts for the latter solution, even though his coffee cost less than her drink, partly because he thinks it is his fault that the cocktail was so awful. I don't know if *Celia* wanted things to go any further.

The contradiction between chivalry and the equality principle could be resolved, were it not for one underlying problem: deciding who pays the bill is not unrelated to what happens next. It is therefore not surprising that the ritual does not supply a specific answer: its role is to create a neutrality that guarantees that both parties are free to decide on the next move. And many people do change their minds when it is time to pay; it all depends on how the conversation went: is this as far as it goes, or do they want to take things a bit further? Curiously enough, a willingness to pay is not a sign of affection. The opposite tends to be true because an old stereotype still influences the way we think about these things.

The problem is that a date can very easily lead to a night of passion. And men are more likely than women to want sex in the short term. It follows that traditional chivalry can still suggest the man is paying for sex. No matter what form the bargain actually takes, the woman is selling her charms or selling her body if she lets the man pay the bill. That this is what is actually going on is much more obvious if the date involves going to a restaurant rather than a café, and more obvious still if it is an expensive restaurant. *Sandra* admits that she has sometimes 'given in to men because they "pulled out all the stops": expensive restaurant, champagne, nightclub. I felt trapped, as though I owed him.' At the opposite extreme, she goes on: 'If you've already decided you're not going to sleep with him when you go on a date, you feel that you're taking advantage, that you're leading him on.'[15] *Steve* feels embarrassed about paying, as he thinks it degrades women. *Frank*, on the other hand, always gets a sexual thrill from paying the bill: 'I can use my money to tell her that I want her.' *Sophie* is an active feminist, but sometimes wishes that the man would pay, even though she has very mixed feelings about this:

> I always offer to pay, but I always hope that he won't let me. If he lets me pay, it doesn't mean we won't become lovers, but it does put me off. It puts me on the defensive and I begin to wonder what he wants from me. When a man buys me a drink, I can lay down my arms and be passive, feminine. I *know* he wants me.

It is therefore difficult to say where chivalry ends and where macho boorishness begins, as the same gesture can be interpreted in different ways. What she wants is a woman's only point of reference. If she really wants to sink into his arms, she is not going to be upset if he pays for the drinks. On the contrary, it is a nice gesture, an invitation, and even a taste of things to come. But if

---

[15] *Sandra* on the Psychologie.com web site.

she is unsure of what she wants, the old stereotype (sex in exchange for money) comes into play and prevents her from enjoying being the object of his attentions. *Anne* even takes the view that a woman should 'always pay' (by which she means pay half the bill) if the man 'wants to go further, and you don't'.

How do we know where we stand, given that all these interpretations are possible, and that they are so contradictory and so ambiguous? The simple answer is that we cannot, and in any case we have other things to think about. The bill proves to be more of a problem that one might think, and that goes some way to explaining why it is better to arrange to meet in a café rather than a restaurant. A café is much more neutral. The fact that so little money is involved means that there is less pressure. No one is going to kick up a fuss over a few euros or the price of an ordinary drink. It may not be very romantic, but the bill issue once more goes to show that those who argue the case for meeting in an ordinary café may well be right.

## Getting it right

The new dating code therefore cannot settle the bill issue in advance. It is too complex, especially as the answer depends on the circumstances and can determine what happens next. It is, on the other hand, difficult to understand why this should be the case. After all, we are talking about concrete little gestures, and these are usually the things that allow us to establish a ritual and to guarantee its neutrality. There is therefore something worrying about the fact that a problem that should have been settled long ago remains undefined. A lot of energy has gone into trying to settle this problem, but to little avail. To make things worse, we often confuse cafés and restaurants, which in fact are very different contexts.

Surfers are constantly reworking and refining the dating code, especially on those specialist sites which, like the etiquette books of old, hand out advice to people who are unsure how to behave and draw up lists of all the things you should and should not do. Even in 1999, when the internet was just beginning to be widely used, there were no fewer than 2,500 sites dedicated to online dating. Since then, their numbers have exploded. Ellen Fein and

Sherrie Schneider, who were two of the earliest 'dating coaches', have written a number of best-sellers. Year after year, they repeat and refine the 'rules' that define a good date. But, as the subtitle of their first book indicates (*Capturing the Heart of Mr Right in Cyberspace*), codes as detailed as this relate to only one aspect of dating. Fein and Schneider are talking about serious commitment of the most traditional kind: this is no laughing matter. All the advice they give makes that very clear. Rule 5 states: 'Wait 24 Hours to Respond [when you are asked for a date].' And Rule 21 states: 'Never Date a Married Man, Even Online'.

Such guides are designed solely for those who are completely lost in the modern dating jungle. No one else would put up with such strict rules, which reflect the personal philosophies of the authors. *Scout* objects to what he sees as their intolerable opinions: 'I like women who do not wait 24 hours before replying to an e-mail. I like women who make the first move.'[16] The imperative rigidity of their advice, which is worthy of the nineteenth century, does not fit in with the personal democratization that is increasingly typical of our day; we all cobble our lives together in our own way and in accordance with our own ideas, values and fancies. We do not really want handbooks that teach us how to behave properly or institutional sites, and the dating code is therefore being reinvented. This goes on very widely in countless chat rooms and forum discussions, and it can be a very creative process. The buzz of chat represents a collective attempt to establish the truth.

A lot of issues are discussed. A lot of thought has gone into this, and guides to action are gradually being drawn up. This is true even when the debate is confused, as we shall see later when we look at the issue of kissing and of when sex becomes a real option. When it comes to paying the bill, on the other hand, the debate is still confused and it is still difficult to disentangle all the mutual interpretations that are on offer. *Alexheart*, for instance, admits to being totally confused: 'I've spent so long looking at all the different views that get aired in forums and chat rooms that I have no idea who should pick up the bill on a first date.'[17]

---

[16] Review of *The Rules for Online Dating*, available at Amazon.com.
[17] From *Spike Seduction*.

The following exchanges illustrate how complex the question can be. They have not been edited and no commentary has been added (it would take too long to discuss every sentence!). In her blog, *Brunette* has raised the issue of who pays the bill. This immediately attracts a lot of comments. One line does appear to emerge and it seems to be the subject of a consensus: that it is in fact an illusion, as it takes us back to the interplay between the traditional roles of men and women, even though this has been officially rejected now that women are asserting their financial independence. Does the issue of the bill show that we are much further away from equality than we thought?

*By Brunette*, 14 March 2009, 16:52
Hi Girls
Whenever I go on a date I ask myself who should pay. I've often had dates with men who pay for the first drink, and I've never had any problem with that. But last week, I thought I should pay for my own. In a sense that's only natural, but at the same time I felt a bit embarrassed and I thought that the guy was a bit tight and I didn't want to see him again. Stupid, I know. They can't pay all the time, but you get used to it and ultimately it's a kind of chivalry, a way of showing they like us. What do you think?

*By Brewenn*, 16 March 2009, 9:54.
Hi
As a general rule men who are in work should pay for the meal or the drink. It's only good manners. But given the times we live in I think that we should offer to pay our share of the bill for a meal from time to time. But for a drink . . . Never. Good luck!!

*By Ganesha*, 16 March 2009, 9:24
Hi
As a rule, a date means going for a drink . . . or a coffee. I think that a man who takes that approach can pay for the drink or the coffee. As for a restaurant meal, it depends on how you feel about these things. As a rule, I've always offered to go Dutch for the first meal, and I've never had to open my purse.
All the best.

*By Girlguide86*, 17 March 2009, 8:19
Hi, on a first date, it's usually the man who pays. If you have a coffee, he can pay; in a restaurant, don't choose the most expensive

things on the menu. That way he knows you're not taking advantage so get your purse out all the same and as a rule he will pay. But if he's come a long distance for this date, I pay. Only natural. If not, and if it's just a drink, it's up to him to pay, and if he can't do that what he's going to give u later? So drop him.

*Drugstorereomance*, 17 March 2009, 18:49
I never pay . . . we girls are high-maintenance: hairdresser, make-up, clothes, bits and bobs . . . it all costs . . . guys aren't . . . if they want to go out with a pretty woman who looks good, they can pay for the privilege !!! . . . And if the date is properly planned, by which I mean a gap of at least a month and a half between getting in touch and going out together . . . that also tells you he's not chasing after anything in a skirt on another site at the same time . . . The men who complain are tight and only after one thing . . . the rest aren't. And a man can always work within his budget when he asks you out . . . we girls aren't asking for the moon . . . LOL.

*By Jeyline0*, 22 March 2009, 12:17
Hmm! I've read the comments, and it depends if you really like him if you want to go on with the relationship then you let him pay, but if he's a wally and talks about nothing but himself and his exes and takes no notice of you all night until you get to the end of the meal and then wants his money's worth careful, you'd be better off paying for your own meal and you'll never see him again . . .

By *Drugstoreromance*, 22 March 2009, 18:39
I don't agree with you, Jeyline, if he's a wally, you have to let him pay, mainly because he already thinks he'll have you for dessert . . . but you have to stay polite right to the very end! There's no point in seeing him again. There's no point in paying if you have to put up with an idiot who talks about his ex . . . I know, I've been there, done that, got the tee shirt. I enjoyed my dessert and my coffee, and I was thinking about something else . . . I go home and as a rule I burst out laughing . . . jerks are the best . . . no danger of falling for a jerk.[18]

---

[18] From *Brunette*'s blog, available at loveconfident.

*Seductor22* does not have all these problems. He's hit upon a great trick. The internet is an Aladdin's cave for those who know how to operate. *Seductor22* has dated 70 women in the space of 4 years. The trick is to meet them all in the same café. It's run by a friend: 'I don't believe it. You're with another girl. Where you do find them all?' The owner was impressed, but he also realized that he could turn this to his own advantage. Eventually, they struck a deal: one free drink for every two he ordered when he was with a girl. So *Seductor22* can make a show of being generous by offering to settle the bill; he is in fact only paying for his own drink.

# 3

# How to Kiss

## A new courtly code

The courtly code celebrated by the poets and troubadours of the twelfth century completely revolutionized the way people felt about their emotions, and it had considerable influence on how their mores evolved over the next few centuries (Kaufmann 2011 [2009]). It had two striking features. The rules of the game were codified to an extraordinary degree. And love was a steady, controlled process punctuated by highly ritualized stages. We live in a society that swears by fun and creativity, but, in both these respects, the 'first drink' ritual is every bit as formalized as courtly love. Detailed accounts of the protocols and conventions involved can easily be found on the net, and the debates that do arise are about how fast we should take things and not about the underlying principle, which everyone (or almost everyone) accepts.

It will be recalled that our love birds certainly did not respect these stages in the online chats they had before they actually met. On the contrary: the keyboard introduced a protective sense of distance that allowed them to say things they would not dare to say otherwise – often directly and crudely. Going on a first date is like starting all over again. They forget about their strange but recent past: they are with someone else now and it is as though this was a new beginning. Because it is so coded and conventional, the ritual of dating is reassuringly neutral, and that makes the transition easier.

What is more to the point, the ritual feels coded and conventional. But, as we saw when we discussed the issue of who pays the bill, it is much less predictable than it might seem. How do you say 'Hello' on a first date? This should not pose any problem, especially in France where the habit of kissing everyone (or almost everyone) on both cheeks when we meet has almost become trendy. But, paradoxically enough, being kissed on the cheek by someone they have never met makes many people feel slightly uncomfortable. It can be seen as a sign of affection rather than a polite gesture, and might suggest that the first stage in the ritual is being rushed. The kiss on the cheek is therefore strictly regulated and is not especially significant in emotional terms. In some cases, two people may say hello without making physical contact, or even by simply shaking hands. When you think of the warmth of their earlier exchanges (and, in many cases, the exchanges to come), the new hint of formality may seem surprising, but it is quite understandable to anyone who understands the function of the ritual.

It is easier to observe this mechanism in cultures where, as in North America, a kiss on the cheek is not the normal form of greeting. *Caro*, who comes from Quebec, still cannot get over the boldness of her date: 'We met for dinner in a little restaurant that I love. When he came in, I closed my book. We introduced ourselves and shook hands. And then he bent over me for a kiss on the cheek. Oh! I think that's so inappropriate at the start of a first date. . . . I couldn't leave him dangling in mid-air like that. After a moment's hesitation, I gave him a quick peck on the cheek. And I very rarely do that, even with people I've known for a long time and that I like.'[1] *Alexine*, on the other hand, thinks that a kiss on the cheek on a first date can be acceptable behaviour in North America, provided that it is very discreet and kept within strict limits: 'You don't want him nibbling your ear!'[2]

Everyone believes that there is a more or less defined code, but everyone interprets it in their own way. We therefore quickly have to adapt the way we behave to what the other person appears to want.

---

[1] From *Caro*'s blog.
[2] From Patho108.com.

## Should you kiss on a first date?

A simple kiss on the cheek to say 'hello' is not the real problem. The real problem is knowing when to give him or her a proper kiss. When can you or should you kiss someone you have just met? On your first date? On the second? Later? The idea that there might be a 'good dating' code is widely discussed in chat rooms and blogs. So much attention is focused on the kissing issue that there is little room on the web for any discussion of all the other ways we try to please our partners and show them affection.[3] There is one obvious exception (when should you have sex?), and it is even more widely discussed than kissing. But let's not get ahead of ourselves.

The net looks to anyone who uses it to be an endless source of information and opinion. Both the information and the opinions obviously differ, but they can, to a greater or lesser extent, be reconciled and pooled. It should be possible to construct a body of knowledge in the same way that we build a tower of bricks. We ought to be able to form an opinion in the same way. Nothing could be further from the truth. A body of knowledge is always a product of the way the data have been collated. Opinions are always expressed in ways that filter the data they convey and organize them into a hierarchy. Anyone who pays any attention to what happens on the net can see that these things are always structured. Chat rooms exist within their own specific contexts. They are visited by specific populations who often speak a language of their own (casual visitors cannot understand a word of what is being said). And different opinions are expressed in different chat rooms. This is as true of discussions of kissing as it is of anything else. Different sites have different answers to the question 'when should you kiss?' Strangely enough, the same question is asked all over the world, even (somewhat subversively) in cultures that are influenced by tradition. And there is always the feeling that the answer is obvious.

---

[3] The issues of privacy, of the interplay of looks and of body language, are widely discussed on more institutional sites, and especially those designed to help younger people meet people of the opposite sex. But unlike the issue of the first kiss, the advice given is not widely discussed.

We are not going on a world tour. Just a couple of jaunts to, for instance, the Muslim world, which is extraordinarily diverse. One of the more sober suggestions to be put forward on *Mac125* ('The Portal for Young Tunisians') comes from *Beautifulcity*: 'I tend to think "why not?" When all's said and done, it's just a kiss.' Others have more daring or provocative suggestions to make. But the Moroccan portal *Yabiladi* adopts a very different tone, even though it too hosts an online dating site and is willing to discuss many different topics. There is a moderator, but other censors claiming to be religious scholars (of uncertain status) also intervene to cool things down by listing all the things that are *haram* ('forbidden'). The question that concerns us (should you kiss on a first date?) is in fact so incongruous that they have difficulty in understanding it. Even so, the question does arise. People make naive but 'daring' suggestions, argue that it should be possible to talk about basic desires whilst remaining within the limits of convention and claim that fleeting emotions can be perfectly sincere. They also talk about personal freedom. But is personal freedom compatible with religion's commandments? *Noujoum* is categorical: 'It is *haram*, and that's all there is to it!' *Lamyâa* disagrees completely: '*Haram*. Oh, come on. It's your life so you can do what you like.' A very angry *Abdel1072* retorts that what *Lamyâa* has just said is itself *haram*: we cannot decide what to do with our lives because the future is in the hands of Allah. *Soussia*, who was the first to raise the issue, goes back to the question of whether you should kiss on a first date: 'I know it's *haram*. You're telling me nothing new there. What I want to know is whether or not you've done it. Because I have. More than once.' Encouraged by this show of bravery, *Souado* throws off all restraint: 'Listen, it's best to kiss on the first date. That way, you at least know if he's any good or not. And if he isn't, you move on to the next one. There's no time to waste.' The guardians of the faith are stupefied, and cannot understand this. 'What a question!' *Marki-59* objects: 'It is *haram*! This is pathetic: there are people here who know that it's *haram* and who do it anyway! Where will it all end? *Allah O Akbar*!!' The more conciliatory *Marimar26* attempts to reconcile these opposites, but simply gets confused: 'To be quite honest, you don't need to know what other people think. Do what feels right to you. God is your judge, not other people! So you have to ask yourself if what you are doing is right or wrong. If the question you are asking applies to you, obviously.' In the end, it is *Jasmine-love* who puts a damper on things. What she has to say has, paradoxically, much more impact than the censors no one listens to. After meeting someone online, she met up with him in a café. 'We talked about this and that . . . He was so kind and so intelligent . . . completely charming . . . I admit that we kissed . . . and it was a long kiss.' And then she got a text message 'telling me it was over, that we had to end it all . . . Why did he do that? I never got any explanation. You won't catch me doing that again.' Kissing on a first date might be imaginable and acceptable if we are carried away by our emotions, but perhaps we shouldn't trust loose cannons who are only interested in pleasure. Having sinned once, *Jasmine-love* has decided to revert to stricter principles and to combine the religious and the profane: 'Rule 1: It really is *haram*. (It's not me saying that, it's my religion). Rule 2: I don't want to rush things.'

## At your own pace

It is not only the followers of traditional religious morality who believe that kissing on a first date means that you are taking things too fast. Surprisingly enough, very young people (and especially girls) tend to say the same thing, albeit in a slightly different way. For many young people, the dream date is something that leads to love and that has nothing to do with the logic of immediate pleasure (Alberoni 1999 [1997]). According to the Adosurf web site, a kiss has to 'come from the heart'. If you are truly in love, it is not a good idea to rush things. Hence the need to respect the various stages of what should be a gradual process. *World-Hate-Misery* sums up the dominant view: 'It depends what you expect from the relationship. And in my view that means no kissing on a first date if you really want the relationship to last.'[4] This leaves the way open for more playful forms of flirtation. The consensus that emerges from the discussions at Adosurf is that dates fall into one of two categories: flirtation or true love. Which leads to the

---

[4] From Adojeunz.

paradoxical conclusion that kissing on a first date is a clear indication that you are not in love. As with many discussions about the
right time to kiss, the initial impression that this is a relatively
simple question gives way to confusion and bewilderment. The
only solution is to turn to a very clear definition of what a date is.

These young people are not representative of the many opinions
and feelings that are expressed on the net. The dominant view is
that a first date has nothing to do with emotional commitment,
and that kissing is therefore allowed.[5] Not surprisingly, it is mainly
men who take this view, as they see it as a way of perpetuating
all the old arguments in favour of sex with no strings attached.
But not all men take this view. Quite apart from the fact that some
men are romantics and are looking for serious commitment, let
us look at the interesting counter-example of *Anadema*. He is a
sort of modern-day online Casanova. He keeps an exhaustive and
extraordinarily detailed record of his many dates. He is always
eager to have sex. But he does not kiss on a first date. He prefers
the good manners of controlled stages and the pleasures of making
steady – but not rapid – progress. *Mistral* takes a very similar view
and explains: 'My way of going about things . . . well, I'd rather
wait for what you call a kiss close, and cultivate the art of
"knowing when she is ready and making her want you" rather
than trying it on on every first date. I find it less crude that way.'[6]
*Anadema* and *Mistral* are, however, special cases (when it comes
to kissing on a first date). Most of the men who surf the web are
somewhat surprised to find that this new way of dating allows
them to express views that they thought had been condemned by
a contemporary society that officially respects women's rights.
Even the most male-chauvinistic pamphlets enjoy a good press and
popular success (see, *inter alia*, Soral 2004). Why deny yourself
the pleasure of a little kiss if you get the chance? When it comes
to pleasure, there is no time to lose. Dating is a sort of competitive
sport or 'game', and the first – and very banal – stage is to get a
kiss as quickly as possible; this is referred to as a 'kiss closer'. The

[5] This view is dominant but not hegemonic. One of the main features of today's
dating scene is its extraordinary diversity. 'The fragmented nature of the dis-
positions and expectations involved in these emotional encounters frustrates
the search for a single image of social evolution' (Lagrange 2003: 15).
[6] From *Spike Seduction*.

next stage (a 'fuck closer') is also becoming very rapid. *Flynn* is quite open about this:

> Being a professional risk-taker, I think you should kiss on a first date because:
>
> - It shows that you are not frightened of being rejected (= reckless).
> - The date obviously went extremely well (if it didn't, you're not very good at this, are you?), so why should she say 'no'?
> - Because you want to, damn it.
> - Who was it who said: 'A woman might forgive a guy who rushes things, but not a guy who misses an opportunity' . . . or words to that effect.[7]

*Xplose* takes the same view:

> - Personally, I think a first date should lead to a fuck close. Why?
> - Because it shows you know what you're after. To be quite honest, if you want to get a kiss close and work your way towards the fuck close, the best thing to do is interrupt her while she is speaking or catch up with her as she walks down the street and go for the kiss close then. You have to let her know that you want something (to kiss her), so you kiss her. That's all there is to it. I hope I'm making myself clear.
> - It doesn't commit you to anything at all. I know from experience that it's a case of the sooner the better when it comes to a kiss close (not to mention a fuck close).
> - It shows you've got balls, that you don't want to waste time going on more dates and that you've already had loads of girls (that's the message you want to get across).[8]

The logic behind this argument is that a first date that does not end with a kiss is a failure. It fails to take into account the fact that going on a date might have other attractions, or that your partner might have other plans. *Flo*, for example, is very impressed by how well the kiss-close specialists do, and therefore gives a

---

[7] From *Spike Seduction*.
[8] From *Spike Seduction*.

very negative account of his evening, even though there was
nothing unpleasant about it.

> First date tonight. Everything was going fine. I made her laugh.
> We were in a café: hot chocolate with whipped cream and rich
> chocolate cake. The atmosphere was very relaxed. Then we left
> the café, wandered around for a bit, then found a dark corner,
> somewhere private. I took her in my arms, and tried . . . Just tried,
> because she offered me her cheek . . . FAILURE.[9]

The few women who are brave enough to visit this site (*Spike
Seduction* is designed primarily for men) have no qualms about
putting down all these competitive men. *Katrin*: 'As a girl, I'm
somewhat surprised at the advice you give on how to steer the
conversation on a first date. I find these remarks very aggressive,
and personally I'd run a mile.' *Sandra* agrees: 'I agree with *Katrin*:
it's very aggressive. Personally, I'd drop the guy there and then. A
woman wants to be treated like a little princess.'[10] Other women
take a different view. And the kiss-close specialists listen carefully
to them. Let's listen to *Anja* for example:

> A man who kisses me on a first date certainly makes me happier
> than a man who doesn't try anything. When a guy gives you a
> peck on the cheek at the end of a first date, I tend to think that's
> a bad sign . . . It makes you wonder and you begin to have doubts
> about going out with him again. I don't know if it's any help to
> you, but I kissed my darling man on our first date. I kissed him
> as we parted, and I have to admit I regretted it . . . I should have
> done it earlier that evening.[11]

## Kino escalation

If women (or at least some of them) agree, why do so many dates
end in failure? The pick-up artists have given this some thought
and have reached the conclusion that it is all a matter of technique.
The guy fails in his attempt to kiss the girl because his technique

---

[9] From *Spike Seduction*.
[10] From *Spike Seduction*.
[11] From Yahoo.com.

is all wrong. . . . It is a mistake to go for a kiss close all at once . . . You have to break the ice and get closer to her . . . You have to work up to it. . . . Unfortunately, the advice they hand out consists of commandments from on high that undermine the self-confidence of their less experienced brethren. Listen to the sad tale of *Seeth*,[12] who is obsessed with the nagging question: how the hell do you go about getting a kiss?

> You see a woman because you want to have a loving relationship with her. You ask her out on a date and she accepts. The date goes well, and everything feels good. Unfortunately, you are not brave enough to kiss her, or to create the opportunity to do so. You think that you can take a rain check, that you'll get closer on your next date and that it will be easier to take the plunge then. Unfortunately, it does not work out like that; ever since that first date, this girl's been very distant and now she contrives not to have time to see you again! So: do you have to kiss her on your first date if you don't want to make a mess of it?[13]

The poor man gets a stinging reply from *Garibaldibaldi*: 'Right, perhaps I should spell out what I think. I think that, given that your date was such a disaster, she must already be having it off with some other guy, just to make her forget about your pitiful performance.'[14] A guy who does not have the knack is a prat. And he can come in for a lot of teasing.

The new pick-up artists are taking over the web. Convinced that they know it all, they hand out their advice from on high, and it is couched in a jargon of their own making. The general idea is that the solution is purely a matter of technique. The movement began in the late 1990s in the United States, thanks mainly to Neil Strauss, the author of the famous *Game* (Strauss 2005). There have been a lot of similar books since then (Strauss 2007, Mystery 2007 and Edwards 2008 are typical examples) and they regularly feature in the best-seller lists. Rival web sites sing the praises of the 'mystery method' or 'kino escalation'. As is so often the case, non-English-speaking countries in Europe have adopted the vocabulary without troubling to translate it. A 'player' who is

---

[12] From Doctissimo.
[13] From Doctissimo.
[14] From Doctissimo.

good at 'the game' has mastered the 'kino escalation' that allows him to get his 'kiss close' quickly and smoothly. This linguistic choice is not insignificant. When they are removed from their context and inserted into a different language, the terms acquire strong 'technical' connotations, rather as though they were fragments of a positive science that is both exact and rule-governed. This also has certain advantages, as these substantialized truths generate nouns. 'Kino' was originally a diminutive adjective derived from kinaesthetics (the sense of the position and movement of parts of the body). 'Kino escalation' simply refers to the way touching can be used as a way to get close to a woman. But, on many sites, 'kinos' take on a life of their own and come to refer to the act of touching. A kino is a form of 'physical contact intended to establish a better connection between two people' (Artdeséduire). And if you know what you are doing, kinos can lead to a kiss close.

There are now countless sites that teach laymen the techniques of kino escalation. They are like modern handbooks for the web-based seducer. The advice they give is sometimes so specific as to be overwhelming, and beginners may find it intimidating. Sometimes they simply tell the novice to be bolder: 'Kino really is one the player's best weapons, but it is not a method that requires a lot of experience. Don't hold back: touch your target without worrying about being shy. Touch her hand, stroke her arm, her hair or her shoulder' (Artdeséduire).

There are also lots of smart alecs in the chat rooms and the blogosphere, and they go on and on about the same theories or rephrase them in their own terms. This is *Keeparise*, to quote one of many examples:

- Rule 1. Use KINOS, but don't act like a fairy. A KINO is an expression of your virility, your self-confidence and your alpha attitude,[15] so you have to be firm, open about what you are

---

[15] *Alpha attitude* is another fashionable piece of jargon. It reifies a truth that is in fact much less stable than the term might suggest. It refers to a charismatic form of self-confidence. Once again, the transformation of adjectives into nouns leads to reification and rigid categories by suggesting, for instance, that the world is divided into alpha males and the rest.

doing and not get uptight. Unlike women, men are not naturally tactile. When a man uses kino techniques, he is trying to seduce the girl. Get used to the idea. A kino begins in the early 'attraction' stage of the seduction process. If you start with kinos, the girl won't think it strange if you suddenly touch her. She knows you are a kino. A kino who is firm and virile from the very beginning is expressing his ALPHA nature and his self-confidence. Don't be afraid of being rejected. Rejection is normal: it means that the sexual tension is there.

- Rule 2. Either use firm, virile kinos, or don't bother. Kinos are gradual. Begin with the upper body, the shoulders for example. Personally, I start the kinos directly: one hand on her hip, and then I push. That's what kino escalation is all about.
- Rule 3. KINO ESCALATION. The theory behind kino escalation is that you gradually make the kinos more intimate by using the push–pull technique. Pull her towards you, then push her away. If you watch women – and especially women who know how to relate to men – you'll see that they know all about push–pull. If you go on with the kino escalation, you'll quite naturally have the girl on your knee, in your arms, and then you're into the rapport phase, and then you can kiss close. No problem.[16]

Given all this esoteric or threatening advice, naive beginners like *Flo* or poor *Seeth* may well be frightened off. And yet there are worse things out there on the net. I am thinking of *Ellimac zero* who claims to be an expert of the 'when to kiss' issue and dreams up nebulous theories and endless kinotic categories (A1, A2, A3 . . . B1 . . . C1 . . . D1 . . .). I will spare the reader his tedious arguments. Fortunately, *Popil* tries to bring him down to earth: 'Hmm. It's a bit up in the air, all this. It's all calculated, analysed, spelled out, classified, packed and ready to go in the post. I'm not saying that it's a load of bollocks, but where's the sincerity in all this? What's all this got to do with the magic of falling in love?'[17] Love is obviously the last thing on a player's mind. Sexual pleasure is all that matters. And the pleasure of the game itself.

---

[16] From Verselejus.
[17] From Doctissimo.

## Chemistry

Because they are so obsessed with the logic of competitiveness, game players tend to forget that they are actually talking about love. They often even forget that it takes two to play this game, and that even kino escalation is not purely a matter of individual technique. *Crazykisstown* has had enough of being told what to do; it's like being back at school. 'If we needed to take lessons to make a success of a first kiss and to get the girl at the first attempt, we'd all be taking A-level biology! Women are complicated, you know.'[18] As we shall see, women, who have not been slow to adapt to this new-style dating, make the same obvious points in online discussions. As *Anja* puts it, nothing will happen unless the girl is 'up for it'. *Blusher* describes the incident he witnessed in a bar;[19] it was, in his view, 'heartbreaking' for the boy. The lad was trying to use body language as best he could but kept coming up against a wall of silence. She just didn't want to know:

> Two young people were having a drink at a table near where I was sitting. Their body language and topics of conversation made it obvious that they'd not known each for very long at all. At one point, the boy – and he was tall and athletic – began to lean over the table, hunching his shoulders so that his hands were close to the girl's. His eyes were shining, and it was obvious that he fancied her. The girl didn't move an inch and was sitting bolt upright on her chair. She kept her hands on the table, next to her coffee cup, one on top of the other like a good girl. There was no need to eavesdrop to know what was going to happen next. Their body language said it all. He was being submissive and was begging her to notice him; she refused to do so, and she was enjoying every moment of it . . .

The unfortunate would-be seducer should have realized that there was no point in going on. She was saying 'no'. She wanted nothing to do with his technique and was not going to play along with him. She did not want to have anything to do with him. She did not like him, and she wanted this to end. More sophisticated advisers warn that men who are overconfident about that tech-

---

[18] From Lepost.
[19] From *Spike Seduction*.

nique can begin to suffer from tunnel vision. The most important
thing is to watch her as she is sending out signals (the way she
looks at him, her body language, the things she says) that encourage him to go on. *Kamal* calls these signals 'kiss openings': they
do more to open the way to a kiss than any aggressive kino escalation.[20] He gives an example:

> Let's take an example. I'm with a girl. Everything's fine for both
> of us. We're getting on like a house on fire. I HAVE TO KISS HER.
> I move closer and closer to her, begin to touch her hair, then stroke
> it . . . If she lets me, so much the better, I can go a little bit further.
> I tell her that she has lovely hair, that it smells fantastic . . . I go
> on touching her hair, sensually, really nicely, the way you'd expect
> from someone with a bit of class, looking into her eyes . . . turning
> on the charm like a real ladykiller. If she responds positively to my
> advances – a shy little 'thank you' – then we're well on our way
> to a kiss.

Easy, you might say! It's easy when the signals are both clear and
positive. But it becomes very difficult when all the signals say 'No',
as they did for *Flo* and poor *Seeth*. And especially when she says
'no' but offers no explanation as to why the answer is 'no'. To
make matters worse, the advice available on the web helps, as
usual, to sustain the illusion that it is all a matter of technique:
all you have to do is learn the language that the signals are speaking, and then everything will be fine. And in many cases, the
famous signals are all the more enigmatic in that the woman is
not sure what to do either. She too is watching the ongoing experiment and trying to pick up signals. She can pick up the signals he
is sending, and they can be clearly positive or clearly negative.
Unless everything is obvious from the outset, she will wait for the
chemistry to begin to work and overcome her doubts.

Chemistry. This is the other recurrent theme in online discussions about first dates. If there is no sudden flash of insight that
makes everything clear, it is assumed that we will intuitively know
what we should do. There is just one problem. Chemistry is not
exactly what we imagine it to be. It is not a mysterious intuitive
process that provides us with reliable directions as to what path

[20] From *Seductionbykamal*.

to take. Whilst it does involve the emotions, it is also a product of specific cognitive processes. As Sofian Beldjerd has clearly demonstrated (2008), it comes into play in specific contexts, or, in other words, when our thoughts are too confused and contradictory to be clarified by conscious and would-be rational thinking. If there is a spark, we have a feeling that everything makes sense. And because it all begins to make sense, if only in rather abstract terms, our feelings cease to be contradictory. It might be objected that this does nothing to change the outcome: chemistry intuitively leads us to do what makes sense to us at a given moment. That is the only thing that matters, and, if it gives us a feeling of inner harmony, so much the better. But we must not forget that chemistry is, in cognitive terms, very complex or even chaotic. The smart alecs describe it in substantialist and positivist terms: 'feeling' is a natural way of perceiving things in the same way that sight tells us what colour something is and that our sense of taste tells us the difference between sweet and sour. It is a sort of gift. It may well be unevenly distributed, but it is a very simple guide to how to act.

Unfortunately, the smart alecs are wrong, and this does not make first dates any easier. Chemistry is the product of complex and shifting mental contents, and it is inherently unstable. As I have demonstrated in a recent book (Kaufmann 2011 [2009]), love has to do with two different emotional logics that are both in competition and contradictory. On the one hand, the only thing that matters is the lived experience that, when it gives us a feeling of well-being, encourages commitment. On the other, in keeping with the great tradition of romanticism, it is the emotional impulse itself that redefines the situation. 'Chemistry' means two very different things for these contradictory logics. It refers to different ways of going about things and has two different outcomes. And when we go on a first date, we often confuse these different forms of love. Chemistry is not an exact science.

We also have to remember that it takes two: we have to want the same things, at least to some extent. If only one of us feels something, this is going to go nowhere. Unfortunately, a couple on a first date are more likely to be on different wavelengths than in perfect harmony. Remember how shocked and offended *Caro* was when her date gave her a peck on the cheek? It was a very different story when she went out with someone else.

The guy was really, really nice. And fit! Tall, long light-brown hair, blue eyes to die for, broad shoulders . . . As for the date, we went for a walk, watched some acrobats, had an ice cream on Grande-Allée. We parted at 11:45. . . . I wasn't brave enough to kiss him on the lips . . . so it was a kiss on the cheek. (*Caro*'s blog)

Sadly, the tall, handsome young man was never heard of again. Not enough kisses for *Caro*; too many kisses for *Nathalie*.

We met for lunch, and I felt completely ridiculous at the idea of sharing a meal with a perfect stranger. It was as though we'd never had all those great discussions or shared all those laughs on the net. When we'd finished our meal, he asked me to walk him back to work and I said I would. On the way, he stopped and kissed me. I had no feelings for him, but I was naive and I was feeling a bit fragile at the time, so I thought to myself, 'Well, why not?'[21]

She soon came to regret that, as he proved to be very clingy. That same day, he introduced her to his workmates as though they were going to get married.

The notion of 'chemistry' raises one last question: if a date is all about intuition and chemistry, all these debates about the right moment for a kiss ought to be meaningless. As *Anja* puts it: 'There's no such thing as the right moment or the wrong moment; just kiss him when you feel like it. And besides, if the girl is up for it, she'll let you know when she's ready.'[22] So why are there so many debates on the web? Because the ritual has to create the impression that it is real and that it can solve most problems. People who are out on a date are in fact thinking along contradictory lines. On the one hand, they believe that they can solve all their problems by using their intuition; on the other, they dream of having a code that tells them what is and what is not acceptable behaviour. The truth of the matter is that a date is a combination of the two. What happens on a date is partly a matter of intuition, partly a matter of following the rules that are laid down by the net. Very few people realize that the two things are contradictory.

---

[21] From a discussion forum.
[22] From Yahoo.

Love is often a contradictory phenomenon involving forces that pull in different directions (Kaufmann 2011 [2009]), and the outcome of a date is inevitably uncertain. Today's rituals (Segalen 2009) cannot work in the same way as those of traditional society, which laid down stable rules about how to behave and enforced them collectively. Individuals have become too autonomous (Singly 2005), and our emotions call the tune. Although intuition has become very important, that does not mean that all the debates about when we should kiss are pointless. The idea that there is a code is essential (even though there is really no such thing). And some parts of the ritual have become very stable. I am thinking especially of the central notion of 'a drink'. 'Having a drink' begins like an old-fashioned ritual: we relate to someone in a very obvious way, and in a very ordinary environment. But once a few words and signals have been exchanged, it is the interplay between what two people want that decides what happens next.

The people who take part in the online debates think they can define rules that apply to everyone, but the way people feel constantly redefines everything. Whilst the rules are not widely respected, they do define a theoretical framework. That framework is constantly changing, and it reflects the mentality of its time. The rules tell us, for instance, when the time is right for a first kiss. And, at the moment, the definition of 'when' is changing very rapidly.

## 'If he's a good kisser . . .'[23]

'When' is not the only thing that matters. 'How' is just as important. I am not talking about the technicalities of kissing. Once more, there are countless sites that give beginners (and the clumsy) extraordinarily detailed advice and make kissing sound like a sport that takes a lot of skill. This set of instructions posted on Adosurf is meant for teenagers:

> You mustn't approach it as though it was an exam or an impossible ordeal. Whether or not it's a good kiss depends on your partner, too. The first thing to remember is that you have to breathe

[23] From *Nina*'s blog on Les Vingtenaires.

through your nose. Gently put your lips close to hers. When they touch and half open, your tongue will find hers. If you want to create a gentle, sensual and romantic effect, start with the tip of your tongue, as though you were asking permission to go a bit further ... then slowly put your tongue, which is still touching hers, inside her mouth (avoid touching her teeth), then start the famous slow rotary movement, first in one direction (as many times as you like), then in the other direction. If you make the first move, she doesn't really have to move her tongue, but she can go in either direction if she likes. You can do anything you like! And you can make it more intense by giving her more tongue and kissing her harder if you want to give her the impression that you want to eat her ... but that's really for harder, passionate kisses. In between these two extremes, there's an extraordinary range of possibilities that reflect how deeply you are in love: passion, excitement ... If you want to avoid getting cramp or want to take a breath, you alternate between taking your tongue out of her mouth, just brushing her lips with yours and then French-kissing her again. She'll probably take over and use her tongue as she sees fit.

Some would say that there is nothing wrong with this. Why not learn how to kiss this way? This is the age of the internet, and when we want to learn how to cook Japanese food or how to sand a floor, we go on the net. Others, like *Popil*, who had his doubts about kino escalation, want to know what all this has to do with love.

For my own part, I will simply say yet again that these debates show how easy it is to reduce everything to technique. It is not that technique does not matter. It does matter, but it is always very personal. It is always specific, and the way our partner sees it varies. That is why the discussion between *RedheadedGirl* and *Nina* is so interesting. This is *Nina*:

A few years ago, I was talking to *RedheadedGirl* about a boy I'd been playing tonsil tennis with. 'Is he a good kisser? Because if he is, it means he's a good fuck.' I frowned: what's she on about? But when you think about it, there's no avoiding the conclusion that my best lover just happened to be my best kisser. Well, there's a surprise.

What do we mean by a good kisser? It's a good question. Each to their own taste, but my own view is a guy who forces his tongue

down my throat or who feels he has to French-kiss me non-stop
for ten minutes is off to a bad start. I like kisses that are half-
gentle, half-rough; I want to feel him tonguing me, but not too
hard. The worst of all is the sort of sloppy kiss that makes me feel
I'm snogging SpongeBob SquarePants. Having said that, it's true
that the 'harder' a guy kisses, the harder he'll be later on (but
there's no need to break my teeth). And vice versa: if he kisses you
like SpongeBob, you know he'll be no good in bed.

Is kissing an essential test, especially if the immediate pleasures
of sex are the first things on your mind? *Nina* tries to get away
from that idea, which suggests that the date is doomed from the
outset. 'But I'm not going to stop right there just because the first
kiss was unsatisfactory. You never know. After all, it seems that
there's an exception to every rule. But when I don't like the way
a guy kisses me, I just sigh inside and hope that he'll make up for
it in other ways. But he never does.' Which would seem to prove
the point.

Nina, however, is an intellectual perfectionist and refuses to be
satisfied with such simple conclusions. Why should there be a
direct link between a kiss and what happens next? Is it purely
physiological? So much so that there might be a link between the
way a guy uses his tongue and the way he uses something else?
Or is it just that first impressions install a filter that alters our
subjective vision for a long time to come:

> Is it really true that a good kisser is a good fuck? A kiss is a very
> important preliminary. It raises the temperature. And if it doesn't,
> I'm not exactly going to feel like going any further. So, are we
> saying that if things get off to a bad start I can make up for lost
> ground later? Can I catch up with him in a race I'm bound to lose?
> A little voice inside my head is beginning to say 'No, I don't like
> the way he kisses'. Why not stop moaning and just assume that it
> has to get better? Or will it all be as bad as this, whatever happens?
> Which came first: the chicken or the egg? Blah blah blah.

The important thing, according to *Nina*, is that we have to be
careful not to rate how men perform in strictly technical terms or
to assume that there is one standard that applies to everyone.
Sexual desire can take many different forms. 'It might actually be
a question of compatibility. As I was saying not so long ago, one

girl's lousy screw is another girl's best lover. It all depends on what you want, and what you expect to get out of sex.' But having expressed all these reservations, she is still very impressed by *RedheadedGirl*'s theory: 'Well, yes, if you like a good hard fuck and you find yourself with a guy who is happier with the starfish position, it's going to be slow and painful.' Carried away by her own enthusiasm, she seems to take her conclusions a little too far: 'It seems only logical that, if it's true that a good kisser is a good fuck, it's hard to see how it can all work out if he's not exactly your idea of the ideal lover.' What does she mean by 'all'? What about love and long-term commitment on a one-to-one basis? A kiss is obviously not a test that tells us what our whole sexual future will be like. But it might tell us what sex will be like in the immediate future. But surely the way we experience a kiss depends on what we are looking for: a brief sexual interlude, or lifelong commitment. I'm sure *Nina* would have a lot to say about that, too.

The issue of emotional commitment is now decided by a gradual process of experimentation. We run tests. On a first date, we are both on trial. Or rather, we test ourselves by looking at what happens between us. Do we feel at ease with each other? Do we really want to go on with this? In my *Premier Matin* (Kaufmann 2002), I look at what happens after a couple's first night together, or during the later stages of the dating process, and at how both partners watch how they react to banal, everyday activities, such as getting out of bed, having a wash, having breakfast... The way we react may come as a surprise and may be embarrassing; it might make us want to run away, or it might tell us that we feel comfortable with our partner. The situation is so complex that we cannot decide anything on a coldly rational basis, and intuition becomes a central part of the decision-making process. We can 'feel' whether we want to leave or stay. What is being intuitively tested is quite clear: do I want to be with this man or this woman? Is this a sustainable relationship, and can it be a source of psychological comfort and pleasure?

We do not experience first dates in the same way because these are not the questions we are trying to answer. The search for pleasure is more important than the quest for well-being. First dates are predominantly a matter of physical desire, sudden impulses and immediate dislikes. We feel adventurous and we can be impulsive. We are not there to explore the possibility of a long-term commitment, and the most likely outcome is that we will end up in bed together. That is why so much hinges upon the first kiss.

We should not, however, oversimplify things. Minor details do influence the way first dates work out, and they are part of the search for pleasure. The slightest upset can completely change the way we experience a first date. Remember *Anadema* and *Celia* in that gloomy café, or the way *Saskia*'s boots hurt her feet. If your feet hurt when you kiss someone, that may well change the way you feel about the kiss. And it might completely alter what happens next.

# 4

# A New Dance

## The revolution in the dating system

Kissing is now an inescapable stage on the road that leads to love. It comes after the preliminary skirmishes and before the closer encounters that will, or so the more desirous candidates hope, come later. The only problem is knowing *when* to kiss. It has not always been this way. One hundred years ago, or just as love matches were beginning to replace arranged marriages (Sohn 1996), only a minority of engaged couples kissed. Most of those who did so belonged to the lower classes; such things were just not done in polite society. Physical relations began, without further ado, with penetration; women regarded sex as a 'conjugal duty' and the idea of pleasure did not come into it. Women were largely responsible for the change (Kaufmann 2011 [2009]), and the role played by kissing was far from negligible. Greatly influenced by the culture of romanticism and the novels they had read, women began to insist on the need for foreplay. Kissing became an activity that existed in its own right and had little to do with the idea of any conjugal commitment. In the first half of the twentieth century, 'the increase in the number of times women kissed men who were neither their first sexual partners nor the men they later married was a sign that the age of sexual modernity had dawned' (Lagrange 2003: 28). In his famous report, Alfred Kinsey (1948) notes, however, that women were still shy about kissing: couples did little more than brush each other's lips.

The revolution in the dating system speeded up this process. All too often, we assume that dating has always been much the same as it is today. For a very long time, couples – and especially middle-class and bourgeois couples – met rarely, on the fringes of society and in secret. Otherwise, lovers' trysts were confined to the pages of romantic novels. Engaged couples were strictly supervised and their families kept a close eye on them. Whilst they were no longer told who they could and could not marry – the importance of love was increasingly recognized – they did have to conform to a strict ritual. There were no exceptions to the rule. A number of authors (Bailey 1988; Fass 1977; Modell 1991) have studied the emergence of what they call 'the dating system' in the America of the 1920s, and it really was revolutionary. The new trend developed mainly amongst an urban youth who both rejected tradition and rebelled against the older generation. It was probably one of the first examples of a new and autonomous youth culture. John Modell emphasizes that it was originally bound up with new forms of music; the new dance halls helped to change the way young people met. Beth Bailey goes so far as to claim that the dating system is one of the keys to understanding the generation gap that was to have such an effect on the whole of the twentieth century.

## What flirting means

In what was still a very prim and proper post-Victorian age, flirtation was characterized mainly by a desire for 'individual satisfaction' (Modell 1991: 67). It was a way of rebelling against the hold of both institutions and tradition. Young people found new ways to flirt by arranging to meet secretly, and alone. It was a way of escaping from the adult world and its conventions. As the restrictions of set gender roles became more relaxed, the dating system 'changed the nature of relations between the sexes' (Lagrange 2003: 51). For a long time, self-control – especially on the part of girls – ensured, however, that things never went too far and that kissing was the main form of sexual activity. For half a century, or until more recent expressions of sexuality emerged, the kiss, as celebrated by Hollywood movies, was the iconic image of love.

Whilst the age at which people first had sexual intercourse fell only slowly over that period, the age at which they had their first kiss fell by three years from 17.5 to 14.1 years for boys, and from 16.6 to 13.6 for girls (Bozon 2008a). Young people had invented a new life phase characterized by the erotic games of flirtation. This was a sign of things to come, and young people were not the only ones to be affected. The dating system completely reorganized interpersonal relations. You had to be attractive if you wanted to be asked out, and that implied displaying and making the most of those personal qualities that are likely to get you noticed. Everyone tried to put themselves on display or to advertise themselves, and that opened up a gap between the pick-up artists and those who were less at ease or less bold with members of the opposite sex. Success (which could be measured by the number of dates you got) was not defined in purely sexual terms. It was also, and perhaps mainly, a way of improving one's self-esteem and intensifying the feeling of being alive. What was once a model for young people is now found throughout society.

Being liked is not, however, enough. You have to give something in return. Especially if you are a girl. As Hugues Lagrange (2003: 49) makes clear when he looks at the role of kisses and cuddles, 'every part of the body has its price'. The boy expects a tangible reward in exchange for taking a girl out on a date (Can he kiss her? Touch her breast? Go a bit further?). The girl is involved in a subtle series of negotiations, and torn between her immediate desire, the need for self-control and the gradual strategy that will ensure that there will be other dates. These are the rules of the game. They are discussed and refined day after day. Boys and girls discuss them in single-sex groups and go into great detail (with, no doubt, some embellishments) as they compare their experiences.

The rules governing these erotic exchanges no longer apply only to young people. And the rules for first dates have changed a lot. There are no more set moves, because the situation can now be interpreted in many different ways. Feminist criticisms of the male-chauvinist archaism have in theory made traditional roles – men make the first move, whilst women haggle over the price – unacceptable. Women are demanding the right to talk openly about sex and to say what they want (and they have proved very good at doing so on the net). There is no longer any need for the old system of haggling, or at least not in a form that pits men against women. Face-to-face dates do, however, reactivate some of the old schemas. Kino escalation is one example: the man uses his tactile skills, but he will get nowhere unless the woman sends out the right signals. Every part of her body corresponds to a stage in the escalation process. It is, however, no longer possible to set the collectively recognized rates of the 1950s. A woman who is in love will sell herself much more dearly than a woman who, like a man, is just looking for sex. If she lets him kiss her, it may be a signal that she does not want things to go any further. It might be an invitation, but it might also be a warning not to expect too much. It can be difficult to tell.

Fortunately, the belief that the chemistry will work gives the impression that we can trust our feelings of the moment with having to worry so much.

## Would you like to dance?

The dating system emerged quite quickly in 1920s' America, and was in synergy with the emergence of a new style of music and the invention of dance halls. At precisely the same time in Europe, the chilly mood and the regression that followed the disasters of the Great War implied a reversion to traditional roles, especially for women (Sohn 1993 [1992]). The fact that Europe and America were so out of step considerably increased the world-wide cultural influence of the United States, and gave the impression that America was a long way ahead of the rest of the world when it came to the evolution of lifestyles. In many respects, things were in fact beginning to change in Europe, too. And the way couples met was one of the things that were changing.

John Modell (1991) notes that the emergence of the dating system was closely bound up with new styles of music and the dance craze. The change in the way couples met was associated with dancing in Europe, too. I invite you to make a detour via this history, which began as early as the nineteenth century, as it has a lot to tell us about the background to today's rituals.

I will not go too far back in time and I will not rewrite the history of dancing, which has been described often enough. The history of the waltz, in particular, is well known. Rémi Hess

(1989) had clearly demonstrated how what was once a scandalous dance was recuperated, codified and institutionalized by polite society until it reached its apotheosis in Vienna. There was still something subversive about it: the couple faced each other and were in close physical contact. As they whirled and twirled, they became dizzy and seemed to be cut off from the world around them. But despite their apparent freedom, the unprecedented rigour of the ritual, which required physical discipline, and the fact that the group was watching ensured that the conventions were observed. The man who asked a woman to dance was being closely watched by her family. It was not the waltz that really revolutionized the way couples met.

The revolution began on the popular fringes of society and during festivals that were alive with giggling couples. Even Brittany's religious festivals were subverted, and countless kisses were quietly stolen on the fringes of the official processions. But it was the local dance that really heightened expectations. The idea that they might meet a girl or boy set young people's hearts beating as they dreamed of the looks, words and caresses that could be quietly exchanged. Families and village communities were well aware of the dangers, and they put a lot of energy into the attempt to define and apply strict rules. Priests were in the forefront of the struggle, and the most radical of them tried to have dances banned altogether. In 1820, one pamphleteer attacked the censors in the name of the villagers who were being 'prevented from dancing', and especially in the name of the girls because 'penitent girls enjoy dancing' (Paul-Louis Courier, cited in Sohn 2003: 95). Unfortunately for the puritanical society in which he lived, it was too late. It could not resist the growing pressure of sexual desire: young people already wanted to have a life of their own and to live for the moment. Long before American music set their feet tapping and made couples dance even faster, dancing made it possible for boys and girls to meet in new ways.

The big change had to do with the new independence of young people. The grip of the family was relaxing, and young people had greater freedom to meet whoever they wanted to meet in purpose-built spaces that were out of public sight. Until then, they had been under constant surveillance. Parents used all their authority to organize whole teams of spies who kept an eye out for the slightest kiss or cuddle: elder brothers, uncles, neighbours, the priest and even the schoolteacher were all recruited. Even the chief of police might be roped in to help. One officer condemned the moral permissiveness of the working class in these terms: 'as a girl, the woman known as "R" worked in the mill and, like all the girls who worked in such places, enjoyed going to dances' (Sohn 2003: 97). Polite society had its balls, but custom dictated that girls were chaperoned by their mothers and were allowed to dance only with their brothers.

By the end of the nineteenth century, all these attempts to mobilize the old forces had become pointless: bodies had been set free, and the pressure of desire was too great (Corbin 1990 [1987]). During the *belle époque*, the number of places where young people could meet increased. A dance was no longer an episodic event associated with a feast or festival. Dances were organized on a regular weekly basis by café owners and innkeepers. New forms of music had emerged, and there were now dance halls. Young people were now free to choose who they met, and where they met. Nothing could hold back the tide.

The days of sedate waltzes and polkas were over. Frenzied rhythms were imported from the Americas: the foxtrot, the charleston, followed by ragtime and boogie-woogie, and then the more sensual sounds of the tango and the samba. The rhythms grew faster and faster, and those of a religious disposition were quite convinced that this was the devil's music. Couples embraced and enjoyed a new physical freedom (Jacotot 2008). Girls sometimes danced with men they did not know, or scarcely knew: dancing with someone did not involve any long-term commitment. The great era of flirtation had begun. The dating system quickly made it more popular, and gave it a generational legitimacy, especially when, as in the United States, the main impetus behind it came from the middle classes who, as always, give the stamp of official recognition to fashionable trends, even though they do not launch them.

## The whole world is a dance hall

A lot of interesting things could be said about this development. First, the dating system gave individuals a new historical opportunity to escape the hold of the group and to obey their impulses of the moment. As we shall see in Part II below, the internet is now accelerating that trend and making it more widespread. But for the moment, I want to talk about the ritual itself. I am not so much interested in its external forms as in the underlying principles behind dancing or, to be more specific, behind asking someone to dance.[1]

One of the major contributions of the so-called Chicago School was to demonstrate how important the 'definition of the situation' is to any individual who has to do something. Social and personal definitions crystallize to produce a single meaning; if they do not, relational mechanisms seize up or lead to confusion. Beggars hold out their hands to ask for money, not to say hello. Police officers stop girls because they are riding their bikes the wrong way down a one-way street, not because they want to chat them up. One of the peculiar features of dancing is that it can be interpreted in contradictory ways. It is what Michel Bozon and François Héran (2006: 62) call an 'ambiguous language'. We think that we are operating on one interpretive register, but we have in fact already slipped into the other. That is why dancing is at once so ambiguous and so powerful. And, to that extent, it is an early example of the uncertainty that surrounds contemporary dating rituals: the ritual is always the same but it can be interpreted in two different ways.

What are the rules of the game? A young man asks a girl to dance. She can say either 'yes' or 'no'. If she says 'yes', the same question arises when the dance comes to an end. If the couple are getting on well and obviously fancy each other, they can extend the experiment with a second and then a third dance. They might hold each other closer, which indicates that they want to go further. When he asked her to dance, there was absolutely no hint of a possible sexual encounter. For official purposes, he asked her to dance and that was all. As with the modern suggestion 'let's

---

[1] It is interesting to note that there is nothing new about the choice of a café as a meeting place. At the beginning of the twentieth century, cafés were the venues for dances, and they were meeting places for gangs of young people in the 1950s. Dance halls have gradually given way to nightclubs where everything revolves around music and the search for a sexual partner. Dancing itself has lost much of its old importance; and, besides, couples no longer dance together.

go for a drink', the very ordinariness of the ritual masked the fact that he might be hoping for a lot more than a dance.

In relational terms, this structure serves two purposes. If she 'knocks him back' after their first dance together, he can avoid the painful side of the situation 'by hiding behind its institutional definition: "I just asked her for a dance" ' (Bozon and Héran 2006: 62). If he can do that, he will suffer no serious psychological damage and his self-esteem will remain intact. He also has a second, and more important, advantage. Asking a girl to dance makes it much easier for the boy to take the initiative. Asking a girl for her hand in marriage used to mean an unshakeable commitment, and the proposal was made with her whole family looking on. In contrast, asking her to dance looks just like a game. For official purposes, there is not a lot at stake: it's just a dance.

Nowadays, it is 'just a drink' and the context is slightly different. There may or may not be music and the couple sit facing each other, with no further physical contact to begin with. But meeting for a drink can be the start of something that quickly leads to a sexual encounter; it can therefore mean that less time is spent on flirting. It is, however, still possible to interpret the situation in two different ways.

At the beginning of the twentieth century, a combination of new forms of music from the Americas and the emergence of the dance hall allowed a new dating system to develop. Young people were therefore able to break down the old institutional disciplines and were free to find new ways of meeting. At the beginning of the twenty-first century, another technical innovation (the internet) speeded up this process and made it more widespread. Dating is no longer an activity that is restricted to young people; it concerns all age groups all over the world. It has become incredibly easy to arrange a date with someone, and a date can easily lead to erotic games that have nothing to do with the idea of any conjugal commitment. It may well be 'just a drink'. But a first drink can easily lead to a second, and a second drink has a very different meaning.

The whole of society has become a huge dance hall where anyone (boy or girl) can ask anyone else (girl or boy) to dance. The underlying principle is that of the dances of old. As in the past, the ritual is reassuring because it is so well-oiled and gives the impression that the rules of the game are clear. But what we make of a date is not predetermined and can vary a great deal, more so than ever now that sex is part of the equation and may even be the most important thing about it. That ambiguity did not matter very much when mere flirting was the order of the day. But now that 'just a drink' can mean more than just a drink, it leads us to confuse love and sex.

# Part II

## Pleasure and Feelings

# 5

# Should You Have Sex on a First Date?

## Wait a while

Sex on a first date? Some will say that I am exaggerating and that the question does not usually arise. Ellen Fein and Sherrie Schneider (2002: 117–25) are very clear about where they stand: their Rule 13 states 'Don't have sex on dates one, two or three'. Although the time-gap between meeting someone online and going to bed with them is, in historical terms, becoming shorter, most people who go on dates do not have sex immediately.[1] A quick visit to a chat room appears to confirm this impression. It was *Jojo28* who raised the topic of 'Do you prefer a girl who waits before she has sex?' For her, this is not an abstract question. *Jojo28* is in a hurry to get an answer. She goes on to explain:

> I've been going out with someone for a week. We've seen each other twice since we got together. Once at his place and once at mine. When he came to my place, we kissed a lot and he put his hand inside my tee-shirt. In the end, he told me he wanted to spend

[1] Within the space of a single generation, the time-gap fell on average from 12 to 6 months for women, and from 10 to 5 months for men (Bozon 2008a). The growing popularity of online dating from the early 2000s onwards has changed this pattern and the time-gap is now much shorter.

the night with me, even though that had not been part of the plan. He was very sweet . . . lots of kisses and cuddles. I don't want to rush things; I'd rather take it slowly. I want to get to know him better, to know if he's serious. I'm spending the night at his place tomorrow. What about you? Do you think you should wait before you have sex with someone when you want the relationship to last?[2]

Several of her online contacts quite understand why she wants a quick answer. *Kantse* warns her: 'If you're going to spend the night at his place, it's more likely than not that you'll end up in bed with him. Good luck.' *Dounette* tries to dissuade her: 'If I were you, I wouldn't spend the night at his place.' She has strong views on the matter: 'I much prefer to wait, even though it can be difficult sometimes. And you never know where you are with sex.' But most of her contacts take a more moderate line and tell her she only has to wait a little bit. *Pariss*: 'I'd wait to make sure he's not just trying to get me into bed. In any case, I'd advise you to wait a while; a week isn't very long.' *Ani* ('To be honest, I did some stupid things when I was younger') suggests waiting for a month. A month seems to be a reasonable length of time. Is this the basis for a consensus? No: *Wgalou* bursts in to say that she can't stand all this talk of timetables. Love is nothing to do with compulsory norms. If there's any chemistry at all involved, you just *know*.

> I've just read the answers and I'm gobsmacked. How you organize your sex life is up to you. If it feels right, and you want to, get on with it! You have to stop worrying about stupid things like this and stop laying down rules when the way you feel should be the only thing that matters. If the chemistry is there, there's nothing wrong with spontaneity. If you ask too many questions, you lose the beauty of spontaneity. And that's so sad.

*Kantse* agrees: 'Do whatever you want and worry about it afterwards.' So, there might not be any need to wait for a month, and she can rely on her intuition. This is a quarrel between those who believe they should work on an intuitive basis, and those who

[2] From a chat room on Yahoo.com.

think they should wait 'a little'. It is unlikely that any clear consensus will emerge. At this point, *Lionofthesouth* interrupts the debate. He is a sexual predator who has somehow strayed into the chat room, and his scream of rage adds to the confusion. 'What! A week and you still haven't slept with him! I don't believe it! It's like listening to a bunch of old women. He made it perfectly clear what he wanted on the second date, when he got his hand inside her tee-shirt. So you know where he'll want to get on the third. It's not rocket science.'

## 'If you want to'

Debates about when to kiss lead to the conclusion that dates fall into one of two categories: if you are really in love, you should be able to wait. The same applies to sex. *Dounette* sums it up perfectly: 'If I want it to last, I wait.' It all depends on what you want: immediate pleasure or long-term commitment. *Lionofthesouth* obviously takes the view that pleasure is the only thing that matters. His bravado is met with protests because the discussion between *Jojo28*, *Dounette* and *Pariss* has led them to a very different conclusion. That much was clear from *Jojo28*'s initial question: 'Do you think you should wait before you have sex with someone if you want the relationship to last?' *Samir*, for example, does not see how there can be any alternative. 'A week isn't enough to be sure you want to be with someone . . . You don't know how they might behave and react in different circumstances. And you have to remember that the whole point of a relationship is not sex but feeling comfortable with someone.'

Those involved in many other chat rooms take the view that feelings are the only thing that matter and simply do not spell out how long anyone should wait. Unfortunately, just what they mean by 'feelings' is not as clear as it might be. When a girl 'feels' that she is in love, it may mean that she has found her soulmate and is ready for a long-term relationship ('I knew he was the one for me'). Or it might simply mean that she wants a physical relationship now ('I felt like it'). As we have seen, these different possibilities lead her to very different conclusions when it comes to deciding how long she should wait. If she feels that things might get serious and that she is in love, she will not have sex on their first date.

'Feeling' is supposed to be more romantic than 'calculating', rather as though it was a magical expression of love itself. Feeling is purer, and more authentic. It is also more modern, whereas the idea of waiting takes us back to the conventions of traditional society. That is often how it is seen in online discussions (*Wgalou*'s contribution is a good example). No one occupies the moral high ground for very long, and it is what they want now that really matters. 'If you want to' becomes the ultimate criterion.

According to *Carole*, all this talk about whether a girl should or should not have sex on a first date is pointless. 'It all depends on the relationship, on the chemistry. If both partners feel all right about it, then there's no problem.'[3] She is trying to find the words to define something that goes beyond desire but she has to fall back on abstractions like 'the quality of the relationship' and 'feeling right'. *Isabelle* does the same and evokes intimacy and trust: 'I have no problem with sex on a first or second date. If there's a feeling of intimacy and trust on the first night, I'm not going to deny myself.'[4] But as the discussions go on, all these euphemisms are soon forgotten about and more physical considerations come to the fore. 'Wanting to' is the only criterion that matters. *Louna*: 'If you want to, and he wants to, why wait?' *Alexine*: 'I did and it was great.' Where's the harm, if it makes you feel good? And, in emotional terms, there can be something powerful, even admirable, about the idea of acting on the spur of the moment. *Amstello*: 'Almost all my girlfriends went to bed with me the night we met. I like the feeling of "we didn't want to be apart."'

A high proportion of people in chat rooms still believe in the 'double standards' principle, and most of them are women. They take the view that, if you think you've found a soulmate, you should wait a little. *Francy*: 'Now that I've reached thirty, I'm beginning to see the difference between serious relationships and sex dates. If I sense that the guy is serious, I wait. If I sense that the guy is just a dick on legs, I make up my mind quickly. I usually say no, unless the lad's fit or unless I'm gagging for it.' Waiting, which was once a rule established by a collective morality, is

[3] From Patho108.com.
[4] Ibid.

increasingly seen as a personal choice, as a signal that says 'I'm not going to rush things because I want a long-term relationship.'

Waiting a bit when you think this is serious, and not because of what other people might think, is something you do because you want to. It proves you are in love and it is an ordeal that comes straight out of the courtly code. But the real change is that we are beginning to have our doubts about over-rigid categories. It is not as though love was so important as to be the complete opposite of pleasure. Is there any reason why a moment's pleasure shouldn't be important in its own right?

## Where's the harm?

Laying down rules about how long we should wait before we have sex looks more and more archaic at a time when it is our emotions that call the tune. Why should our desires of the moment always be mediocre or shameful? In some cases, simply satisfying basic needs can have unexpected consequences: 'a sex date can lead to a real relationship' (*Pkoi*). Many of the people who are involved in these debates find all the talk of feelings disturbing, and would rather turn them into openly moral discussions about what is right and what is wrong. Not in terms of collective rules that have to be obeyed, but in terms of a personal quest for happiness. 'I don't take the "no sex on a first date" line. I don't always have sex on a first date either. It all depends on how I feel, on the context. I have had beautiful relationships that began as sex dates, and I've also met a lot of bastards. Takes all sorts to make a world' (*Helen*). Being able to tolerate other people's choices is what matters. That way, we can satisfy our needs of the moment and find our own path to happiness.

'What are we frightened of? Of being judged? Do we have to do as others do before we can feel good about ourselves? Are we doing this because it's what is expected of us, or because we want to? Our sex lives are no one's business but ours.' Should you have sex on a first date? Do you have to? *Pouepine* has all sorts of obsessional doubts about how normal her life is, and about the meaning of right and wrong. 'The sex thing really worries me, that's true. Am I normal, or am I going off the rails? Am I becoming decadent? I never know how to describe where I stand. What's right, and what's immoral?' The rights and wrongs of sex on a first date might not constitute the height of philosophical debate, but they can involve some unexpectedly profound questions. This is often the case when it comes to defining moral principles, and such debates often focus on what we wrongly assume to be insignificant events. The 'sex on a first date' question raises the issue of the role played by sex in the symbolic architecture of our society as a whole, and of how physical pleasure relates to commitment.

As our moral points of reference are defined in terms of our personal happiness rather than in terms of what other people might think, our definitions of what is wrong tend to come down to a few awkward situations. They may well be problematic, but they are also limited in scope and they can easily be defined. They include the physical dangers that might arise from a date that goes wrong, and the feelings of psychological discomfort ('I feel empty' . . . 'I can't take any more' . . . 'I feel sick') that overwhelm us when having more and more dates no longer gives our lives meaning, or when it no longer gives us the pleasure we anticipated. Then there is the feeling of weariness when we finally become bored with having the same old conversation over a drink, or when a partner thinks only of his or her own pleasure,[5] or does not even have the good manners to be polite when the time comes to say goodbye. There are moments of terrible loneliness, and then there are the difficulties that arise when we try to commit ourselves on a more long-term basis.

Compared with these problems (which are much less forbidding than the terrible taboos of the past), there seem to be no limits to the freedoms we now enjoy. Now that lovers are out of society's sight, surely they are doing no one any harm when they have a good time? Arranging a date on the net is something special because it allows us to escape from a normal world that is often too quick to pass judgement. When we are in the café or the

---

[5] It is hard to strike a balance between the pleasure we give and the pleasure we receive. Women, for instance, enjoy cunnilingus whereas men enjoy fellatio. 'Whilst both men and women say that giving their partners oral pleasure is their favourite sexual practice, both sexes would rather receive than give' (Ferrand, Bajos and Andro 2008: 365). This average imbalance can become extreme in situations where a selfish partner becomes dominant.

bedroom, mutual pleasure is the only thing that matters, and it has little to do with established conventions. The only morality that counts is the morality of shared well-being and mutual enjoyment. And then normal life resumes. In its heyday, romantics dreamed of how overpowering emotions could let us leave our ordinary lives behind (Kaufmann 2011 [2009]). Online dating makes that simple, physical, immediate and almost banal. There is none of the pain and heartbreak associated with great emotional upheavals. There are none of the dangers of a journey from which no one will return. It's a whirl of pleasure, but it only lasts for a while. Rather like a dance.

## 'Slag!'

Yes, but . . . men and women are not equal in this respect. Far from it. And so long as men and women are not equal, it will be difficult to enjoy these enchanted interludes (or what look like enchanted interludes). It is easy for men. When they visit chat rooms, they strongly advocate an easy-going philosophy: adopt a relaxed attitude towards physical pleasure and everything will be all right. Women should, they say, stop denying themselves. But there are other men who are saying something very different (or perhaps they are the same men at different times). They want long-term relationships and are looking for soulmates. And suddenly the liberated woman they used to dream of begins to look very different. She becomes repulsive, and they begin to call her all the names under the sun. She is a 'whore' and a 'slag'. 'If you have sex on a first date, you're a slag' (*Etienne*). All the weight of centuries and centuries of sexual taboos that discriminate against women bears down upon women who thought they were liberated and were equal to men in every respect.

*Stephie82* is outraged. 'What are you on about, *Etienne*? Why is it always the woman who's the slag? And what about the guy who has sex on a first date? What's he, a saint? It takes two, you know . . . So men can and women can't? Bizarre.' It is not in fact all that bizarre. Or only if you believe all the promises and all the talk of equality and forget about the mechanisms that still restrict women's freedom. History weighs heavily upon our mentalities, and a few declarations of principle are not going to make it disappear as though by magic. And there are many areas in which discrimination against women is still official, especially when it comes to sex. Isabelle Clair (2008) points out that girls living on the big estates on the fringes of French cities are kept under close surveillance. Nothing they do or say goes unnoticed, and this affects the way their image and reputation change. A girl who is 'immodestly' dressed, or who has been seen in public once too often, must be a slag. It goes without saying that any suggestion of a sexual relationship is all it takes for the abuse to begin. 'When they pass her in the street, they're like "Oh, she's the one who's sleeping with What's-his-name." They'll even write it on the wall for everyone to see' (*Karine*, cited in Clair 2008: 79).

Women are still under surveillance. Even when they think they have escaped it, they find it much more difficult than men to make their sexual freedom a reality. For a while, *Cheapgirl* enjoyed behaving like a boy: she was just out for sex and never thought about tomorrow. 'It was a time when my girlfriends got a good laugh out of me, but it seems I got a reputation for being a man-eater.' She wants to turn over a new leaf, but 'the image stuck, and I can't shrug it off.' The net is the only thing that can destroy such old ideas and give her a new identity. On the net, she is free to define herself as she sees fit. But as soon as she arranges a 'real life' date, she has to meet a whole series of conditions if she wishes to retain that freedom. A big town where there is no danger of running into someone she knows; a café she does not usually go to. If she does arrange a date, she has to be careful not to let slip anything that might get back to anyone she knows in real life. And she certainly has to avoid choosing anyone who is not a complete stranger. 'Choose someone who is not part of your usual circle of acquaintances (no colleagues, no mates, no local shop-keepers).' Groups talk, gossip, make and break reputations. That is what they always do and what they have always done. And they are always quick to condemn a woman who steps out of line in the name of an unspoken moral code that is never spelled out in so many words.

And to make matters worse, men have a lot to say about their sexual conquests when they are in male company, even though women complain that they tend to say nothing once they are in a steady relationship. There are, once again, historical reasons for this; ever since the age of chivalry, seducing a woman has been seen as equivalent to a feat of arms (Kaufmann 2011 [2009]). Men do not just boast (and exaggerate) about their conquests; they go into great detail and recommend 'easy lays' to their friends. This does nothing to improve the image of the woman concerned. *Lisa1236* is still reeling from the shock, and that is why she began the debate on the web. 'Do you think a girl who goes to bed with a man on their first date is a slag? Because that's what I did, and the guy's friends insulted me. Because the idiot told them everything.' *Stephie82* advises her: 'Sleep with whoever you like, when you like, just makes sure he knows how to keep his mouth shut.' Coming after *Etienne*'s imprecations, this is all too much for *Lisa*. She screams in rage at the unfairness of a system that allows boys to do almost anything and gives girls much less freedom (I have to apologize for her language). 'What a shit! And what about him? I suppose it's nothing to do with him! He's the slag, not you. What a lousy bastard the guy is. Shooting his mouth off like that! All you have to do is tell his mates that he's got a tiny prick and that he's lousy in bed into the bargain.'

## A world apart

Within the space of only a few years, the internet has revolutionized the way we establish social relations. Especially when it comes to sex and love. As I said earlier, the process began with flirtations in dance halls. It helped to create new spaces where young people could escape from moral injunctions and where the hold of the group was more relaxed. Here, couples could share moments of emotional intensity in a sort of little place of their own that was cut off from the rest of the world. The internet suddenly speeded up this process. The possibilities are endless. What takes place on the net helps to create 'a world apart' where the normal rules of the game simply do not apply (Miller and Slater 2000). This world is not as virtual as it is sometimes said to be, but it is 'liquid' because it has been freed from the constraints that bind us to territories, groups and established conventions.[6] Women

---

[6] Society is, as Zygmunt Bauman (2003) puts it, becoming more and more 'liquid', nowhere more so than on the internet, which is doing a lot to make society more fluid.

can at last behave just like men and define their own moral standards without having to worry about the stereotypes. In societies
that are inscribed within territories, the social bond is based upon
established proximities. No two individuals are ever alone. As
soon as they come into contact (unless they can really succeed in
hiding away), they bring with them a whole process of social
penetration (Merkle and Richardson 2000: 189), with its codes,
its values, its disapproving looks and its judgements. Merkle and
Richardson note that one of the main characteristics of the internet is that it inverts this pattern: two individuals can come into
contact on a private, subjective basis. They are not subject to any
restraints or a priori definitions.

At least in theory. The 'world apart', known as the internet, is,
as Miller and Slater so nicely put it (2000: 16), a source of 'potential freedoms' rather than real or complete freedom. Indeed, there
are also a lot of damning stereotypes around, and they can be
used in violent ways. This is quite understandable if we remember
that chat rooms and forums are the places where an attempt is
being made to define a common moral code, and that cannot be
done unless people have strongly held convictions. In her blog,
*Helen* dreams of a sexuality that is free, serene, relaxed and
happy, and then she is surprised to find herself being described as
a 'nympho' or a 'slut'. But she answered back, mocking her critics
and recruiting an army of women to back her up. *Etienne* (who
called *Lisa* a slag) found himself isolated and came in for a lot of
mockery and vilification. He may reoffend in a different chat
room, or he may change his online name, but the girls have put
across the idea of equality for the duration of the debate. Potential
freedom has become real freedom.

Like the dance halls where couples went to flirt, the internet is
a world apart that exists parallel to worlds in which socialization
is more firmly grounded. There are many links between these
worlds. And those links are both unbroken and complex. As the
'world apart' becomes more cut off, it becomes easier to escape
the stereotypes. Changing behaviours is now a real possibility. All
groups that are subject to rigid conventions, that suffer discrimination or that simply dream of enjoying greater freedom therefore
use the internet as soon as they have the technical ability to do
so. They attempt to expand the spaces in which they are free to
express themselves. Young people (and even children: Dumesnil

2004), women and members of cultural communities that are ruled by tradition all use the internet.

## Cafés elsewhere

The example of Muslim women is especially interesting. The strictest interpretations of the Koran forbid women from meeting men or being seen in public places. But the internet is not a public place. In a country like France, the number of self-professed Muslims (men and women) who use dating sites is proportionately higher than the number of non-believers or Christians (Bozon 2008b). 'Using the Internet allows them to expand the geographical range of their dating activities and to avoid parental controls on what they are doing' (Bozon 2008b: 286).

Let us take a little trip to Cairo. In the 1970s, there were lots of secret corners in Cairo where couples could meet without being seen ('We used to make love in the al-Ornan gardens', admits Farida: Heshmat 2006: 18). The increasingly repressive climate means that there now many fewer such places: 'Cairo has chased away its lovers.' As a result, 'The Internet has become *the* place to look for love . . . and it allows strangers to meet away from the traditional circuits' (Heshmat 2006: 20). The same applies to Kuwait, where online connections allow women to 'talk to men without worrying about the social consequences' (Wheeler 2006: 146). It is the same in Jordan. Let's take a closer look at that country in the company of Laura Pearl Kaya (2009) as she visits students at Yarmouk University in Irbid.

On campus, the shops, the fast food outlets, the westernized décor and the open spaces create an atmosphere that encourages people to mingle and express themselves, especially as the students are a long way from home. But that is not enough: someone is keeping an eye on them and opportunities are limited. *Mona*, a 19-year-old Palestinian, spends two hours a day in chat rooms. A lot of men are in love with her on the net.

*Mona* enters this new world via the Rishrush internet café. Irbid has so many internet cafés that it should be in the *Guinness Book of Records*. In 2004, or ten years after Jordan went online, there were no fewer than 104 and they were very popular with both male and female students. A lot of interesting things go on in

Irbid's internet cafés. By logging on, anyone can enter a parallel universe where very different rules apply to the game of love. But there is more to it than that. The cafés are amongst the very few places where boys and girls can get close to one another and where the sexes are not segregated.[7] More important still, the net can be used to organize discreet excursions into real life: subtle techniques can be used to contact potential boyfriends or girlfriends who are actually in the café. The messages then become more and more flirtatious and sexualized as the girls and boys exchange sly glances.

One day when she was in the Rishrush, *Mona* replied to *Romeo* with realizing that he was actually in the café. They had fun trying to identify each other, and he told her that she was a romantic.

[7] Pearl Kaya stresses that cafés with big windows overlooking the street tend to be frequented mainly by men. Women prefer cafés that are clearly separated from public space.

# 6

# Sex as Leisure Activity

---

## When sex broke free from feelings

Not everyone in the Rishrush café is a romantic Romeo. A lot of the men who go there also visit porn web sites. Sex is everywhere in Irbid, just as it is all over the world. By 'sex', I mean sex for the sake of it, sex with no strings attached, free of all constraints, the kind of sex that breaks the old taboos in its search for new thrills. The 'parallel universe' known as the internet has freed us from the few restraints that remained and has turned a minor fire into a global conflagration.

If we really want to understand what is happening today, we have to go back in time. Without wishing to rewrite the entire history of sexuality, which is both rich and complex, we have to go back, if only briefly, to that crucial moment towards the end of the twentieth century when sex began to be completely divorced from feeling on a long-term basis. We had already experienced sexual revolutions, some exuberant, some joyous and some violent. But what happened at the end of the nineteenth century had much more far-reaching implications because it almost completely divorced sex from feeling for a long time to come.

It all began with romanticism. Romanticism swept through Europe like a wildfire. We tend to forget that it began as a political and philosophical protest movement, and that it was not confined to the private sphere. The idea behind romanticism was the need to reject the narrow horizons of the new mercantile civilization that was beginning to dominate intellectual life. It was hoped that something exciting but indefinable would make it possible to transcend the mediocrity of ordinary life. Romanticism represented an extreme and desperate attempt to reach the sublime. It failed to achieve its highest ambitions and went astray as it degenerated into an obsession with the mysterious, suicidal depression or an archaic conservatism. But, thanks to an astonishing reversal of fortune, its failure led to the discovery of the soft, pastel colours of 'romance', or of the sentimental, courtly romanticism that is now so familiar to us (Kaufmann 2011 [2009]).

All over Europe, theologians, philosophers and doctors attacked the radical side of early romanticism. This mad passion was a threat to the new bourgeois order, and nothing good could come of it. To hell with all these medieval mists and this suicidal melancholy! Love had to be part of ordinary life once more. The new science of psychiatry even decreed, with all the medical authority it could muster, that mad love was madness pure and simple and that lovesickness was just a sickness. And the sick had to be cared for. After all the torments of romanticism, all polite society wanted to do was get back to normal.

When romanticism began to take the form of romance, and when women appropriated the romantic novel in order to expand the space available to them (partly because politics and economic were out of bounds to them), the literary avant-garde turned against all this mawkishness. Love became a 'ridiculous, naive feeling' (Bologne 1998: 177). A surprising reversal meant that it was 'sentimentality rather than sexuality that seemed indecent' (Bologne 1998: 177). Thanks to the scientific discovery of nature in the course of the eighteenth and nineteenth century, and the development of the great materialist philosophies, feeling, and especially the feeling of love, came to be seen as mere illusion. Indeed, according to Schopenhauer, love was the enemy. Freud's ideas brought about a revolution by demonstrating that sexuality governs the unconscious mind and structures our personalities in childhood, but he saw love as a form of alienation, and passion as a form of infantile regression that bordered on psychosis (Alberoni 1994 [1992]). Jean-Claude Bologne (1998: 170) remarks that, according to the father of psychoanalysis, our feelings are nothing more than 'egoism, projected fantasies and reflections of our drives. It is not a woman that we love; it is an ideal image that transcends any individual woman and that relates to our mothers.' The more political philosophers who came after Freud, and who used sexuality to promote social liberation, were made in the same ideological mould. Wilhelm Reich, the apostle of the sexual revolution (1951 [1936]), called upon us to cast off all our chains, including our emotional chains. Herbert Marcuse (1955) contrived never to talk about love, even though all his work deals with the links between sex and society.[1]

When the sexual revolution did break out in the 1960s, it was first and foremost a force that broke the chains of a puritanical society that had reached an impasse and in which individuals were not really at ease with themselves. The inevitable process that speeded up when the pill allowed women to divorce physical pleasure from the need to reproduce inevitably prepared the ground for something with which we are now only too familiar: the world of sexuality became completely autonomous. It had been set free from the shackles of sentimentality and was ready to establish new rules for the social game. But what were those rules?

## Sexual liberation: how do things really stand?

The hedonistic 1960s utopias of unbridled pleasure and the sexual emancipation movement triggered a real 'Dionysian earthquake' whose aftershocks are still being felt in both quantitative and qualitative terms. 'A process that is both inevitable and dangerous continues to accelerate' (Lipovetsky 2005: 306). Sex has changed. It has become autonomous and self-sufficient. A relationship is an investment like any other: 'you put in time, money . . . hoping that [you will] be repaid – with profit' (Bauman 2003: 12). Books with titles like the *Secrets of an Expert: How to Drive Your Man Wild with Pleasure* and *All You Need to Know about Fellatio* regularly feature in the best-seller lists. And the number of web sites with a pornographic content is mind-boggling. As for the actual content, there seem to be no limits. Hard porn has become commonplace, and all tastes are catered for, with sites devoted to gang-bangs,

---

[1] This link is made by Anthony Giddens (1992).

bondage, double and triple penetration, S&M, scatology, golden showers, sex toys, wife-swapping . . . The list is endless. Even children are at it. When they go on the net, the most popular entries into search engines, after 'You Tube' and 'Google', are 'sex' and 'porn'.[2]

Major studies of changes in sexual practices, however, lead to much more moderate conclusions. Behaviours are certainly becoming more varied and diversified, and that reveals a growing interest in sexuality, but they are changing relatively slowly and have little to do with the explosion of sexualized material in the media. And the most transgressive practices are in fact still confined to the margins. 'Social reality does not reflect the flood of "hyper-sexuality" that we see in the media' (Lipovetsky 2005: 307). Lipovetsky explains why the two things are out of step; sexual pleasure is all very well but most people are looking for a long-term relationship. We dream of enjoying a very simple form of happiness, and we see the cocoon of domestic bliss as an antidote to the aggressive poisons of modern life. The media are not, on the other hand, lying to us. Both these things are real. Society is obsessed with the search for pleasure, has a taste for adventure and is interested in new and more intense sensations, but it also needs the stability and reassurance that encourage us to avoid risk-taking and not to go too far. That is why current developments appear to be so contradictory. Major changes do take place when conditions are right (think of the ease with which we can contact people online and commonplace rituals of the real-life dates that follow). At the same time, day-to-day life can still be very conservative and humdrum (having sex on a regular basis can just be a matter of routine).

There is another reason why the two things are out of step. Historically, sexual liberation took the form of permanent transgression. The old fetters of Puritanism had to be broken and taboos had to be violated before bodies could be set free. Some still take this transgressive view and go in search of new taboos to violate. They always want to go further and to be even more transgressive, even though there are fewer and fewer taboos to break (there are some exceptions, such as child protection). Hence

---

[2] Figures from Symantec parental controls (2009), based on 3.5 million requests.

the fascination of perverse forms of sexuality. These have become very trendy, even a source of social distinction. De Sade is back in fashion. But it has to be remembered that the defining feature of the 'divine marquis' was the utter selfishness of his quest for pleasure. He was quite prepared to kill his victims if murder could give him personal pleasure. There is something anti-human about the 'tyranny of pleasure' (Guillebaud 1998).

Are we really attracted to this kind of sexuality? Some people are, but even they stay within certain limits: if they do not, perversion can easily lead to murder. But most people are not. Although it is the shocking imagery that gets all the media attention, the real sexual revolution that we are living through is taking place elsewhere. It is much more discreet, much more far-reaching and, paradoxically, much more subversive.

The real sexual revolution has little to do with violence, transgression or egotism. Most people are looking for the sweet pleasures of happiness (which does not mean that they are not intensely exciting), and they want to enjoy them with someone else. It is obvious from the ads on the dating sites that this is what they want. Those who are interested only in sex dating (and most of them men) are not quite sure what to make of this development and it makes them look tacky because it represents a return to the libertine tradition that has always preferred Montaigne to de Sade, and Casanova to Don Juan. The revolution we are going through is making sexuality an everyday matter for all of us, no matter whether we are living with a partner or involved in the dating game (not that the two things are mutually exclusive). It marks an irrevocable break with the very long history in which sex was to some extent bound up with a disturbing symbolic world full of dark forces that was attractive precisely because it was so disturbing. Sex now tends to be something that is simple, normal and pleasant. It has become a sort of leisure activity. And instant gratification with no strings attached can be a very pleasant experience. 'I'm not patient, madam. When I go to the cake shop I buy something to eat now, not in three years' time. That's just the way it is. Men are worse than macaroons. I have to eat them there and then.'[3] You can just treat yourself to sex. It does you good and it

---

[3] From *Pinklady*'s blog.

does no one any harm. Many people who go on dates actually plan their dates in the same way that they plan the rest of the weekend. They might even choose between meeting someone 'for a drink', reading a novel and going to the theatre. And as for people who are in settled relationships, holidays often add a new spice to their lovemaking: two treats for the price of one.

## Feeling good together

Leisure activities themselves have become part of the broader trend that leads us to look for well-being and minor pleasures. Society has become very harsh on us in psychological terms, and we all need to be pampered and cosseted. We need to enjoy more and more pleasant sensations by escaping, however briefly, into an alternative life. We want to have showers that never end (and not only for purely hygienic reasons), perfumed massages and lots of chocolate. The pleasures of sex certainly add a frisson of adventure to dating, but the underlying desire for well-being is becoming more explicit and more common.

The desire for well-being has a long history of its own. There is no room here to describe how relationships were socially engineered to make them a potential source of happiness (Kaufmann 2011 [2009]). At first, this development reflected the influence of theologians and moralists; then, in the twentieth century, the same role was played by psychologists and marriage guidance counsellors. Let me simply note that sex was for a long time seen as a problem. Copulation, which was tolerated because of the need to reproduce the species in the biological sense, was surrounded by all sorts of taboos to prevent it from degenerating into 'fornication', a generic term referring to all kinds of over-indulgent pleasures, feelings and activities. Women had to perform their 'duty' as coldly and unemotionally as possible (husbands were more easily forgiven because they were by temperament more 'red-blooded'). Ideally, they should be thinking of something else (preferably God) and, if they could not stop themselves from feeling some lascivious emotions, they could atone for their sins in the confessional. It took hundreds of years of bedroom struggles before women could assert their right to pleasure and before institutions were forced to accept that the very nature of love had changed.

Long before Kinsey (and then Masters and Johnson), sexologists like Havelock Ellis were showing the way at the beginning of the twentieth century: a healthy attitude towards sexuality could be a source of both pleasure and well-being (Robinson 1976). The sexologists were gradually replaced by all kinds of institutions, including highly prestigious bodies such as the World Health Organization. In 1974, the WHO officially accepted that reproductive sexuality and the search for pleasure are two different things. It recognized the legitimacy and importance of pleasure as a source of well-being and even health (Giami 2005). The imperative to have a good time has become part of the political equation.

The 'sexual well-being' that was institutionalized in this way was obviously defined with legitimate couples in mind. The WHO had no intention of opening the floodgates to an unbridled libertinism. But that is in fact what happened, especially when the internet gave it a new and powerful impetus. This is because online sexuality is far removed from the violence and danger of extreme transgression. It paves the way for a 'well-tempered hedonism', and suggests that there might be other forms of well-being. 'I want us to be very naughty,' *Nick* tells his partner of the moment. He then adds 'but the important thing is that everything has to be fluid, simple and hot'. At another point, he speaks of wanting to enjoy 'cool moments that are sensual, tender and funny'. 'Sometimes, we can even talk about tenderness, and that's fine' (cited in Wingrove 2008: 47, 45, 59).

This is the truly subversive dimension of what we are living through: the paradoxical banality of a new leisure activity that is exciting but risk is free. It is not just about sex in the physical sense of the term. It is also about discovering how two worlds can coincide, if only on a temporary basis. It is about being gentle, about having a good time and about paying attention to what the other person wants. And this is the big difference between the new sexuality and the search for casual sex, which can be a purely selfish quest for pleasure (it may involve lying to partners and then rejecting them without any explanation). The new sexuality, in contrast, is based upon openness, togetherness and, above all, curiosity about other people. So no two dates are ever the same.

It is only obsessional pick-up artists who believe that all dates are the same and that sex is the only thing that matters.

I have already mentioned *Anadema* and all his conquests. He describes and catalogues them in remarkable detail.[4] A quick visit to his site might give the impression that he is an old-fashioned womanizer who is proud of his conquests and boasts about them. He is, in fact, fascinated by the possibility of discovering someone else's private world. That is why, as we saw earlier, he does not expect sex on a first date and why he does not even try to kiss the girl. As in the code of courtly love, he prefers to stick to a gradual ritual that allows him to enjoy the foreplay all the more. He does not want it to last for ever, and he is not interested in long-term relationships. He wants to experience something intense which demands no commitment. There is no obligation to wait either; if it feels right, it can all happen very quickly.

> Last Thursday, early in the evening I was feeling really frustrated after some misadventures that I don't want to talk about, I suddenly felt a furious urge to find a quick date. I wanted to have sex, a bit of tenderness, a good date. I was feeling adventurous. I looked at the list of women who were online to choose some young ladies who were, judging by their photos, pretty with ads that made them sound interesting.

' "A", 27, quite charming, slim, pretty smile' made him want to join in her game. The way he phrased his response says a lot about the new search for 'sexual well-being'. Let's take the time to go into some detail. Within a few minutes they were in a chat room, and trying to explain what they wanted.

## The story of *A*

> 'Could you be a bit clearer about what you're looking for?'
> 'In an ideal world I'd be looking for a woman to spend the rest of my life with. But given that I'm an Epicurean, I'm also interested in having some nice dates. I like girls with personality and a sense

---

[4] The latest address for *Anadema's Story* (*Blog sur mes rencontres amoureuses sur Internet*) is www.anadema.fr.

of humour. What about you? What are you looking for? I've a simple suggestion to make: when you feel like it, let's have a drink one night. If the chemistry is right, we can have a cuddle and see what happens. What you do think?'

'I'd like to have a serious relationship, even though I know it's difficult. For the time being . . . Why not go on dates? You never know, I might meet the man of my dreams.'

'You might end up having a nice time. It might change your life for the better.'

'Yeah, but it's a bit difficult when a guy comes to see you just to get you into bed.'

'Well, I'm not just looking for sex. I want to have a nice time with someone . . . togetherness . . . sensual . . . we can have a laugh . . . share our secrets . . . When it's just about sex, it all gets a bit boring, to my mind.'

'Well, it is better. If there's more to it than that.'

'I think you're cute, and your ad made me smile. I'm suggesting we spend some time together, and take it from there. Simple as that.'

'OK, no problems.'

'When do you want to meet? Tonight if you like (just give me time to finish eating!)'

'Why not?'

Five minutes later – the sound of her voice is not too off-putting – they arrange on the phone to meet at ten. That gives *Anadema* time to finish his meal. Everything can seem incredibly simple when sex is just a leisure activity.

*Anadema* was a little disappointed at first. Her working-class accent is stronger than it sounded on the phone. She talks too much. Her sweatshirt is too big. And to make matters worse, she has a dog. 'Doggie says hello by trying to sink his fangs into my jeans. *A.* asks me to sit with her on the sofa. The dog seems to like me. He gets up onto the sofa and tries to lick my face, puts his head on my knee and leaves a big wet mark on my jeans.' *Anadema* is beginning to think that this 'over-rapid' date was not such a good idea, but he decides to see it through to the end. 'We start hugging and kissing, and then we want to go further. We get up to go into her bedroom. She closes the door, which means we don't have to have the dog on the bed beside us (I'd draw the line at that). Because he doesn't want to be left out, he scratches at the door and whines non-stop.'

### Post-coitus

As usual, *Anadema* describes everything he did and thought with a sense of detail worthy of an entomologist. There is something that intrigues him:

> She's affectionate, and not especially passive but I soon begin to sense that we have something of a problem. Very early on, she quite spontaneously makes the first move and we're in the 69 position, but I can tell she's not really enjoying it. So while we're catching our breath, I ask her why she's done that so rarely (because that's what she told me). She's a bit shy about it but tells me that she doesn't really like giving blow jobs. Oh, shit! I'm gobsmacked. I try to get things back to normal by telling her that we're here to have a good time and that no one has to do anything they don't want to do. She seems to like that and relaxes a bit. We start making out again, and we're both getting turned on. But she often says nothing when I try to ask her if she likes this or that . . . A bit later, I let her put a condom on me and get on top. I can sense that she's more comfortable that way; she's in control and she can move the way she likes to move. We do all sorts of things and change positions. I can tell she wants me to be very gentle and to reassure her.
>
> She seems rather surprised when I spontaneously tell her I'd like to spend the night with her. She's dated guys who've left at three or four in the morning after they've had sex, and seems to think that's normal.

*Anadema* thinks that spending the night with a woman is a nice thing to do, and that it makes the whole experience more complete. She's quite willing but insists that they have to share the bed with the dog.

They say goodbye in the morning. It's all over: that's what they agreed. *Anadema* could quite easily have left it at that, and would have done so if all he wanted was to move on to the next conquest, now that he'd got what he wanted. But he wanted to know what it was that hadn't been quite right. And perhaps he just wanted to help. It hadn't been as good as he'd expected, but he also sensed that it hadn't exactly been the perfect night for *A*. either. It hadn't been what they'd been hoping for. The next day, he got in touch

online and asked her why she hadn't felt more at ease, and if she'd
rather have just . . .

> 'We could just as easily have spent the night cuddling without
> having sex, you know. That would have been nice too, and you
> might have felt more comfortable.'
> 'I'd have liked it if we'd just cuddled, but I thought you wanted
> to have sex and I didn't want to disappoint you.'
> 'You should have said so. I wouldn't have objected. I told you,
> the whole point was for both of us to have a nice time.'
> 'I didn't dare!'
> 'I wanted us to have a good time together. Both of us, not just
> me!'

She gradually came out with it. A number of bad experiences
(including a near-rape) had left her with certain inhibitions and a
poor image of male desire.

> I hope I've at least given her some new reference points, and a
> better idea of what she's entitled to expect from a man, even if it
> is just for one night: tenderness, sensitivity, attentiveness, respect
> and the feeling that she can trust a man without having to do
> anything she doesn't want to do. If that helps to change the way
> she sees relations between men and women, so much the better.

*Anadema* is not, on the other hand, cut out to be a long-term
therapist. He may well visit dating sites in a spirit of curiosity and
even altruism, but he is there to enjoy himself and not to make
the world a better place. And from that point of view, he did not
get much out of his date. He can only conclude that he rushed
things, and that he didn't wait long enough to tell if there was any
real chemistry between them – the kind that guarantees that
they'll both have a nice time. That is also why he prefers not to
get too carried away on a first date.

# 7

# The Game

## The games people play

The picture I am painting is, perhaps, too idyllic. The new world
of dating does not always revolve around the search for moments
of shared pleasure and well-being. Not everything about it is
sweetness and light. Things can go wrong. Accidents happen, and
people have all sorts of unhappy experiences. Dates can turn out
badly. *Thirty-Something* was stood up at the Sofitel in Marseilles.
Poor *Seeth* didn't even get a kiss. *A.* made love even though she
did not want to. Two people who arrange to go out on a date
sometimes have very different expectations, and that can lead to
misunderstandings. Arrangements can be cancelled – suddenly
and in sneaky ways (often by a short text). All these things can
have serious psychological repercussions When there are no limits
and no real purpose, transgression for the sake of it can turn an
erotic encounter into selfish pornography, and the emphasis on
physical pleasure can turn the other person into an object and
'help to do away with the very idea that he or she is a human
being', as Michela Marzano (2007: 13) puts it.

As sex as leisure activity becomes more widespread, it creates
a vast space in which anything is possible. Although they are all
inspired by the idea of pleasure and play, these activities can take
very different forms. The search for a moment of shared well-
being is only one of many possibilities. It is certainly the most

interesting and, although it may not seem it, it is the most subversive in that it offers an alternative to exclusive or permanent relationships. But it is not the only possibility. Some men view sex in terms that have more to do with 'game playing' in a much stricter sense. We have already seen the way in which they use seduction techniques to get to a fuck close as quickly as possible. *Mistral* calls this 'the game spirit: seduction as sport as well as game'. This form of competitive sport takes up even more energy than the search for pleasure, even if we define pleasure in purely selfish terms. It does involve a form of self-transcendence, but it does not bring us any closer to the person we love or desire. Self-transcendence for its own sake can, like any form of competition, become a form of self-destruction.

In a sense, there is nothing new about this. There is a long tradition of male rivalry when it comes to female conquests. As I have already said, the rules were first codified during the age of courtly love when the gallant knight had to survive a series of ordeals in order to win his lady's heart. Today's sexual predators take a less chivalrous view but they are still motivated by the spirit of competitiveness. They compete to conquer as many women as possible in order to prove something to themselves and to have something to boast about to their friends and rivals.

## A popular sport

There is also something new about 'the game', and that is not just because our lives are, to an increasing extent, ruled by competition (Duret 2009). The players who write blogs and contribute to forums differ from the predators of old in several respects. First, the net itself provides a legitimate arena for competitive sports. As in any competition, all the players talk about themselves, rate their respective techniques and laugh at less skilful players. The game has become a sort of official sport that already has its champions and its specialist media, whereas the predators of old operated in secret or had only a very small audience. The other big change also has to do with the internet. It has become incredibly easy to arrange one sex date after another. Playing the game once took a cunning that only the boldest could acquire, but anyone can join in now, even if they are shy or clumsy. It has become accessible, even commonplace. The skill no longer consists in knowing how to join in the game but in developing an interesting technique and the ability to tell stories about it. The other difference is that this really is a game. It is difficult to tell where the virtual ends and where reality begins. Like a video game, it begins on a screen but it ends with a very physical encounter between two bodies. This is not always pleasant or easy for the player's partner, who may not appreciate being unceremoniously discarded. That happens a lot. But from the player's point of view, the exciting thing about the game is that something that used to be so complicated has become so disconcertingly easy. That improves his self-confidence and life feels good. Playing the game is fun.

The basic principle is non-commitment. The player does not want to be tied down: he wants a series of quick dates that lead nowhere. And he has no time to lose. If a woman 'doesn't know what she wants, move on to another' is *Angel*'s advice. *Great Master* is even more ruthless:

> As little chat as possible, either online or on the phone. The goal is to get a date by the second e-mail, the third at a pinch. If it's all just chat, get the hell out of it, and remember that your e-mails should be very, very short. No point to all this smooth talk or chat that goes on for hours.

If a relationship does seem to be dragging on, the level of self-control has to be very high: 'The last thing you want is for her to get serious and become too emotionally involved' (Wingrove 2008: 26). According to *Nick*, 'If you want variety in your fucking, you have to fuck a lot of women' (Wingrove 2008: 27). He explains that the point is 'not to focus on anyone in particular, but to stay alert all the time and to keep your eyes and ears open' (Wingrove 2008: 25). And then you have to develop a technique that works. 'When I say it has to work, I mean it has to get girls into my bed as quickly as possible and with as little fuss as possible' (Wingrove 2008: 24). If they can manage to devote all their energies to the game (it helps to have an undemanding job), the best players can clock up some spectacular scores, but over a limited period of time in most cases (it's the same for any sportsman: overdoing it leads to physical or mental exhaustion). In

*Nick*'s case, it went on for a year 'When everything was going well, there were unbelievable times when I was sleeping with five or six girls a week . . . meeting others . . . and knowing I'd get them into bed before long' (Wingrove 2008: 26).

## Statistics

Many of the men (and now the women) who play the game are scrupulous about keeping an accurate record of their conquests, and thus perpetuate and extend the male tradition of 'notches on the bedpost'.

> I'm a bit like everyone when it comes to my old lovers. Yeah . . . well, yeah. I don't keep count, you know . . . counting the notches on your bedpost isn't exactly cool, is it? I'm a terrible liar . . . and what about you? No . . . straight up, you mean you don't keep count? You don't even try to keep a little list? (*Grenobleboy*)

When *Grenobleboy* asked this question, everyone or almost everyone (including the girls) told him that they kept count too. *Seriousgirl* had forgotten one or two. *Livepen* admits that she only counts the ones 'who've actually come inside me'. *Chulie* is a little shy about these things, and somewhat sentimental: 'It's not as though I keep count . . . but all my exes counted . . . hard to forget them.'

*Grenobleboy* self-deprecatingly describes himself as 'a serial loser' compared with the game's champions. But he still keeps count, and is not shy about posting his score in his blog:

> Good loser that I am, I've not had hundreds of women, so I keep count of them all. Might as well. There's nothing wrong with that . . . the government doesn't have a monopoly on the use of inflated statistics. . . . Until now, I've been able to rely on my memory to recall all their names: three Emilys, Laetitia . . . I may have been a late developer when it comes to my love life, but I still find it a bit difficult to remember all their names . . . in what order . . . and then there's all the details (colour of hair, perfume, cup size . . .). Fortunately, technology has come to the rescue of my failing memory.

He is right to say that technology is important. It has completely changed the way we keep count. Things that were once kept secret – or written in a little black book that was kept under lock and key – are now both anonymous and all over the media. The media supplies tools that do more than allow us to preserve our memories. They also allow the champions to produce sophisticated statistics using Excel tables, graphs and pie charts. Anyone can analyse their performance in impressively scientific terms. *Nick*, for instance, has worked out that on 77.7 per cent of his dates (note the concern for detail!), he got the girl into bed on the first day. 55.5 per cent were 'one-offs', three were 'frigid'; two were 'not too great'; eight were 'hot' and two were 'atomic'.

Appearances can be deceptive. When it comes to statistics, showing off is a secondary consideration. People keep statistics for purposes of their own, and they may not always really want to. Anyone who signs up to a dating site is asked to fill out a profile, and that in itself encourages users to categorise themselves. As they begin to explore further afield, keeping a record of potential dating partners quickly becomes indispensable. They have to be sorted, classified and chosen. Keeping the archives up to date is just as important. It makes it less likely that one person will be confused with another, and makes it easier to remember facts that might come in useful later when we want to draw up a proper balance sheet.

## Some portraits

All this talk of figures can be deceptive. Most gamers talk about numbers when they realize that using the internet makes it easier to get lots of dates rather than having to put a lot of effort into getting a few dates, as they used to do. *Seductor22* (whose arrangement with the café owner means that he gets a second drink free) has had 70 dates. 'If you've never tried online dating, that might sound impossible, but for anyone familiar with dating sites and the womanizers who hang out there, it doesn't sound like a lot.' As it becomes easier to get dates, the number seems to be less important. The point of statistics is not so much the ability to boast (the number of notches on the bedpost) as the ability to learn from them in order to improve your performance and, more surprisingly still, in order to understand what is going on. Players who have been plunged into a new world in which dates come fast and furious want to learn from their every encounter, no matter how minor. They remember their casual partners because they remind them of some significant event, an attitude or a physical feeling. Codifying things makes it easier to understand them. We have already seen how this works when we looked at the café scene. And what happens after the second drink is just as predictable: a little verbal foreplay . . . undressing . . . lovemaking . . . favourite position. Paradoxically, it is the way we do the same things with different people that really brings out both the differences between individual partners and the things they have in common. It also makes it much easier to paint their portraits because the support is always the same.

This is, perhaps, the most surprising thing about players: they have no rivals when it comes to the art of portraiture. They combine a concern for detail (which can be quite shameless) with breathtakingly bold judgements (the anonymity of the net allows them to say things they would never dare to say elsewhere). It takes only a few lines to describe the basics: figure, clothes (and underwear), sexual behaviour and desires, psychological profile . . . Archiving software sometimes make it possible to compare and rate different partners. *Nick*, for instance, rates his partners on a four-to-one-star scale, using criteria such as 'face', 'chemistry' and 'erotic performance'. But no matter how broad-brush they may be, the dominant feature of such portraits is that they focus on sex, in the anatomical sense of the term. Nothing is overlooked: firmness of buttocks, pubic aesthetics, vaginal details. Like the inspectors who write for restaurant guides, players are unsparingly critical of the ways their partners go about doing things. It has to be said that as they have more and more dates – and find them more easily – the competition heats up. This applies to men and women alike. The man or woman who quietly goes about things the way they always have done will soon be pilloried or even denounced on the net. Remember the sad tale of *Soft Toffee*. In many cases, the ratings system is very specific. Men seem to have a real obsession with fellatio and, more generally, the attention that is paid to their pride and joy. Their descriptions of women are mercilessly detailed and, if the commentaries are anything to go by, they rarely rate them better than 'satisfactory'. *Divine Prick* is therefore all the more excited about finding a woman who knows how to handle his precious tackle properly:

> Quiet week. Last Friday, I went to bed with that girl. She's not that good in bed, but she sucks beautifully. To tell the truth, she's really good at giving hand jobs. At some point in her life, she must have met a man who had the patience to teach her how a cock works. Night after night, on the sofa, in the shower . . . he taught her how to find the right rhythm, just how tight to grip a cock. And a proper ejaculation to round it all off. When you think about it, not a lot of women ever meet a hand-job guru. Most of them handle your cock as though they were trying at all costs to squeeze the last drops out of a bottle of shampoo that's almost empty.[1]

## Disgust and cynicism

As the game becomes easier, some players seek out more difficult challenges. *BoySesamplay* is proud of his achievements:

> My greatest internet success was arranging three dates in one day. The first date was at 11:00, then I arranged the second for 15:00 and the last was in the evening at 22:00. To make things even better, the third one was just near my place. I have friends who use dating sites and get dates like me, but I don't know anyone else who's managed three romantic dates in one day. It means synchronizing things, and it means that you have to be fairly successful.[2]

It's even worse for *Divine Prick*; he doesn't know what to do next to keep himself entertained. For a while, he developed a 'new obsession: finding and shagging a woman born on the same day as me'. On another occasion, he broke the habit of a lifetime and introduced a waiting period of – for reasons best known to himself – 23 days:

[1] From *Divine Prick*'s blog.
[2] From plan-drague.net (discussion forum).

Day 23 of the search: make contact and ask her out. She says yes. Day 23 of the search + 3 hours: fucked her on the edge of her sofa. Day 23 of the search + 4 hours; she sucked me off in the kitchen. The earth didn't move when I penetrated her, and my heart didn't stop when I ejaculated in her mouth. I'll have to think of something else.

The game is not without its dangers. As it becomes banal and more and more repetitive, the level of satisfaction falls steadily: the surprise element has gone. The player feels less involved. He goes on playing because that is what he is used to, but his heart is no longer in it. Having too many partners can even gradually lead to a loss of desire or even – horror of horrors – interfere with his ability to express his triumphant virility in a physical way. *Divine Prick* is forced to reach this sad conclusion:

> For the moment, I'm not even in control of my own prick. It's getting more and more temperamental. I can screw a girl two or three times, and then it goes on strike. I blame my penis because it's easier that way. It's always to blame. It's almost as though it wasn't part of me. The real problem is obviously inside my head; my brain's turned to mush after all these years of non-stop sex. So what do I do now to get it up?

Sex addiction follows the same pattern as any other addiction (Giddens 1992): the less the external element (drug or sex) fulfils the addict's life, the more he increases the dose or the more often he shoots up. The law of diminishing returns is beginning to make its effects felt. As *Divine Prick* so rightly says, the real problem is inside his head. To be more specific, the real problem is that sex has lost its meaning, and therefore its interest. He does not even have the energy to do it with any conviction. 'I've always more or less believed in making as little effort as possible, but I was still capable of making the extra effort that allowed me to go all the way. The ways things are going, some girl really will have to put my hand on her pussy to get me started. And I'm not even sure about that.'

Being addicted to the game does not usually have such serious consequences. In most cases, it simply leads to compulsive behaviour (more and more dates that are not really enjoyable, mental and physical fatigue, and less and less involvement with actual partners). It also desocializes players. No matter how many fleeting dates they have, addicts cannot establish any real bonds, become cut off from the people they usually mix with and can end up very isolated. *Madamprevaricates* has not had a date for a while and is beginning to experience withdrawal symptoms:

> What I hadn't foreseen or expected is the unbearable feeling of loneliness. It's not as though the atmosphere was aggressive or tense. It's just freezing cold. I dream of warm, welcoming arms. Of being close to someone and of being wanted. Of having someone to share things with. And after three weeks of doing without, I'm beginning to feel this will never end.[3]

Despite all these problems, they go on playing the game. As *Nick* puts it, all these problems and doubts 'vanish as soon as I get my hand inside a new pair of knickers'. The lack of emotional involvement does, however, have other effects. Unlike *Anadema* or even *Nick*, many players concentrate only on their own pleasure and on the game itself, and are very careful not to become emotionally involved because there is a real danger that this would lead to a commitment that would put an end to the game. There is a high price to be paid for this indifference: players end up taking a gloomy and cynical view of people, and especially women. *Pokerman* has reached a bleak conclusion: 'Don't you think that more and more of the girls on the dating sites are ugly, stupid or psychologically unstable?' They can also develop a crudely direct and utilitarian attitude, rather as though the net had become a vast hypermarket where anyone can ask for anything and everything. *Goodlookingkid* has asked *Pinklady*:

> 'Don't suppose you have a mate who'd be up for a threesome, do you?' 'Excuse me! I'll have you know I'm woman enough for any man!' And I love my girlfriends, so I really don't want to see them screwing in front of me. Even (especially) if I'm having it off at the same time. No problem. He suggests going to a swingers' club. No, no and no again. And I sense that, the more I say no, the more it turns him on.

And then her surprise turns to anger: 'At the time, I was just a bit put out, but then I began to lose it. I'm not a whore, you know,

---

[3] From *Madamprevaricates*'s blog.

so you can't buy my sexual expertise. And I'm not the kind of girl you sleep with because no one else is available. You should see my visitors' book: I'm a sex goddess.' Conclusion 'File under *connard*.'

Most women take a very dim view of this kind of selfish behaviour, and conclude that men will never change. And they have a word for men who are guilty of being so ill-mannered: they are *connards*.[4]

[4] The reader will have to excuse me for using an expression that is vulgar, but there is no avoiding it as women often use it to describe a typical male attitude. In everyday usage a *connard* is 'stupid bastard' or 'prick'. In the world of online dating, the word is used to describe this kind of male attitude. I cannot stop myself from noting that there is an ironic twist to this recent linguistic development, as the term's etymology shows that the insult makes reference to the female genitals. Thanks to an astonishing reversion to the original meaning, a *connard* is a man who is obsessed with cunt (*con*, from the Latin *cunnus*).

# 8

# The 'LoveSex' Imbroglio

## A little love

The game can be fun for a while. But it very quickly proves to have its limitations. It ceases to be enjoyable, and that rather defeats the purpose. The all-pervasive cynicism and utilitarianism eventually sicken anyone who has any sense of human decency. When the players become too cold and too detached, nothing good can come of it. If casual sex is to be a game, it has to be based upon new rules that make at least some allowance for love. Or, if 'love' sounds too off-putting, for affection, for a little attentiveness to our partners, given that they are human beings and not just sex objects. If that could be done, the micro-adventure of online dating could mean something very different: it could be a way of escaping from ordinary life, of discovering an unknown world (as in the romantic tradition) or enjoying an idyll for two that takes us far away from the world in which we usually live.

The problem is that introducing an element of affection means that a date might easily develop into a long-term relationship. And it is only possible to view sex as a leisure activity if it is clear that sex has nothing to do with a loving commitment. How can we be closely involved with – or even passionate about – someone for just a short time? How can we forget about them and just move on to someone else? That is the difficulty we have to resolve, and it is unusual for it to be resolved successfully. Given the uncertain nature of modern dating, it is hard to grasp the difference between pleasure and emotion. One partner may be interested only in sex whilst the other is already dreaming of love. The man who thought he was concerned only with his own pleasure may find that he does have feelings for his partner. A date – which might involve nothing more than a drink, but which might also end up in bed – is always something of an emotional roller-coaster because we are trying to strike a balance between pleasure and feeling. Especially when we try to introduce a modicum of affection.

We therefore have to find ways of being affectionate that do not involve any commitment, discover ways of loving on a strictly temporary basis. We have to try to stabilize the new emotional space created by online dating. To put it a different way: given that both love and sex are always problematic because they are so unpredictable, we should try to create a new entity that stabilizes them and combines them in specific ways: 'LoveSex'. The stakes are high. If such an entity did emerge, it might provide a serious alternative to a traditional relationship. The utopia of an open, loving world in which there are no lies and no conflict might become a reality (Chaumier 2004). A compulsive addiction to the game can, in contrast, desocialize and isolate the players. *Nick, Anadema* and many others suggest that there should be a clear arrangement (a sort of fixed-term contract) from the very beginning. Guaranteed non-involvement on the part of both partners would, in emotional terms, allow them to commit themselves without holding anything back. We would play at loving each other, but only for a given period of time. Sadly, such emotional self-control is very difficult to achieve. And it is the preserve of players who go on so many dates that they do not really have time to settle into a relationship. Even *Nick* has had his doubts from time to time: he has found himself falling in love. *Madamprevaricates* explains why telling herself she cannot fall in love is so painful: 'Because falling in love would be unmanageable. Because I don't want my life to be even more complicated than it already is. I have enough problems as it is. Despite all the things I want to give you, I must not fall in love with you.'

No, there really has to be something else. *Marion* suggests the idea of a network of 'fuck friends' or 'friends with benefits'.

## FWBs

It all began with an outrageous text posted by *Ra7or*, an extreme predator who makes no secret of the contempt and indifference he feels for the women he preys upon. This is one of his more restrained contributions to the debate:

> Appearances to the contrary notwithstanding, a sex date is not like a normal girl. She has the same physical attributes, but offers none of the psychological benefits. As definitions can get a bit confused these days, let *Ra7or* put you on the straight and narrow and help you set limits for your sex dates. That way, you can avoid all the borderline cases. You have to know what you want . . . So you want a relationship that's cool, someone to talk to, someone to cuddle up to? Then you need a proper girlfriend. One who has arms as well as a vagina. It's called 'investment', in the sense of 'investing your own money'. That's a drag when she's a sex date but it's cool when she's a girl friend. The sense of getting a return on your investment is very different. You feel satisfied, at ease with the world. Having a drink or a meal with a sex date is obviously a waste of money.[1]

Always anxious to save a few pennies, *Ra7or* advises against taking a sex date home. He'd rather 'spend the night at her place. You can split as soon as you're finished. You've got better things to do and, if that makes her miserable, well at least it's her cornflakes she's eating and not yours. It's a win–win, as we say in the business.'

*Marion* was shocked, but not by the idea of sex dating itself: 'Let's not play the innocent, we've all been there. Get your stuff together, a kiss on the top of the head (just to show there's no hard feelings), and off you go home to feed the cat.' She is, on the other hand, shocked by the intolerable cynicism, the complete lack of any human feeling or affection: 'It's the crude way you describe it. Just a fuck and that's all there is to it. No cuddles. You don't even stay the night with her. No public displays of emotion. It's like if you never get to know anyone, you'll never become attached.' *Marion* comes up with a very different theory. Her Friend with Benefits (or Regular Sex Dates) system is primarily emotional, and

---

[1] From *Ra7or*'s blog.

is very different from *Ra7or*'s distant, cynical utilitarianism. 'It's subtle. You have a cuddle, sleep together, have a chat if you feel like it, have a laugh, talk things over, go for a couple of drinks.' Well-being plus pleasure, so to speak. But a Friend with Benefits is also someone you see on a regular basis over a period of time (so you have time to talk as well). And that simply means that certain rules have to be respected:

> There's a 'but'. This kind of guy does not really become part of your life, and you don't become part of his. You know his friends by name, by reputation, but that's all. All you know is that you're enjoying what you are doing now. It's the same for him. You don't need to see each other every night, and you don't want to lay down conditions (daily contact, exclusivity, fidelity . . .) when everything is fine the way it is.

A Friend with Benefits is part of a closed system that does not overlap with the other circles in which she moves. He is always available when she wants him (assuming he's not otherwise engaged), but he is always somewhere else, in a world apart, in the anonymity of the net. He is a hybrid, a cross between a friend and a lover, but his defining characteristic is that he is not the only man for her. 'He's fine as he is, but he's not really the man for me. Either he's not the one, or this isn't the right moment.'[2]

The point about a network of multiple relationships that exist in parallel is to try to resolve the difficult problem of how to combine affection with a refusal to commit.

## A new relationship with exes?

Having a network of FWBs completely changes the way people see their exes. The traditional view is that it is difficult to have anything to do with an ex, always assuming that we do not immediately forget them. The longer the relationship, the more we remember the break-up and the quarrels: we tend to remember all the things that went wrong. *Kilgore* has the feeling that his ex-partners try to avoid him when he bumps into them.[3] *Neoryuki*

[2] *Marion*'s blog.
[3] presence-pc.com (discussion forum).

saw his ex-girlfriend only once. 'I just wanted to get my own back, and to piss off the guy she dropped me for. And, while I was at it, to do all the things I'd never dared to do, either because I was shy or because I respected her too much.' Including sadistic things: 'A good night of really dirty, bestial sex, the sort of things you would never do with someone you have feelings for.'[4]

Having a network of friend-lovers changes all this. There are no more bad memories of a break-up because there was no break-up. People sometimes drift away or gradually vanish from your life. But there are no tears and no recriminations. If we do remain in contact, it is all very low-key, relaxed and almost conspiratorial. This has no impact on the people we know on an official basis: an ex exists in a world apart.

Young people are more likely to have occasional contact with their exes (Beltzer and Bozon 2008). As this practice becomes more common, it gets easier to relate to them because there is no longer something special about an ex. So people who were lovers can go on being friends. It would, however, be a mistake to paint too rosy a picture, especially when we are talking about calling upon a supply of emotional resources as and when we feel like it. That is not so simple.

First, it requires good organizational skills. You have to be able to manage a portfolio of parallel relationships, skilfully and without making any mistakes. And, as the network expands, and the longer it lasts, the more difficult it becomes to sever the links and to forget everything when you think you have met the man or woman of your dreams. *Loveapple* cannot bring herself to give up everything:

When it was just about sex, I was quite open about it. 'Bye, guys. I'm turning over a new leaf. Loveapple has left for pastures new. Thanks for everything: it was great. Goodbye and good luck.' I can't do that with Miko. I'm on very affectionate terms with him. I tell him everything. It's like having a girlfriend (but he's no girl, and I should know). Just a kiss or a little cuddle when things are not going too well. We're not cheating on anyone by staying friends.[5]

[4] presence-pc.com.
[5] Yahoo.com discussion forum.

The most difficult thing about a network is that it means rec-
onciling the demands of different FWBs. You suddenly feel the
need to be with someone, so you send a text to X or Y to see if
they are free. But what you actually need is not always entirely
clear (sex, affection, someone who is just there for you . . .). And
besides, X or Y may not want the same thing, always assuming
that they are not otherwise engaged. Just arranging a time and
place may prove difficult. *Missreckless* was having a good time
with her girlfriends and did not want to leave them:

> It was only 21:30 when he got in touch: 'Where are you? Are you
> still coming over? I'm knackered and I need to go to bed, so hurry
> up.' He was beginning to get on my nerves. It was only 21:30 and
> I was fine where I was. When I finally got to his place at about
> 23:00, he was already in bed and half asleep. I lay down beside
> him and he complained that I'd had too much to drink (well, it
> was someone's birthday so of course we'd had a drink). He was
> being a real pain. I didn't even want to kiss him. Fell asleep and
> left at 6:30. I don't think I'll be seeing him again. Not ever.[6]

The biggest mismatch between expectations inevitably has to
do with the issue of love, and there is always a danger that it will
crop up again. The more at ease you are with someone in the
network, the greater the temptation to breach the contract that
says 'No commitment . . . Never, never.' *Missreckless*, who has
said her final goodbyes to her early-to-bed ex, secretly dreams of
coming to a new arrangement with someone else and of getting
him to move in with her. She has a nice house, but it is beginning
to feel lonely:

> The problem with an easy-going relationship – 'no future', 'just a
> bit of fun' is that, after a bottle (or two) or wine in such a nice
> setting, you begin to say to yourself that it would be nice to have
> someone to share it with. Except that you can't call him because
> it's not 'serious'. I'd love to be able to ring him, to send him a text
> saying that I miss him, that I'd like to have him in my bed tonight,
> wake up with him tomorrow, sunbathe by the swimming pool,
> have lunch in the shade on the terrace and then have a Provençal
> 'siesta' . . . but I can't do that. I have to be stoical because I don't
> want to frighten him off.

[6] *Missreckless*'s blog.

## People still want long-term relationships

Online dating has had a major impact on the landscape of love, and things will never be the same again. A new space in which sex can be seen as a leisure activity has been created, and it is protected by the anonymity of this 'world apart'; we can plan a night of passion in the same way that we plan a trip to the cinema or a meal out. It now looks as though Charles Fourier's utopia of a 'new world of love' might be undermining the importance of monogamy (Fourier 1967 [1816]).[7] The steady worldwide rise in the number of people who remain single also indicates that a long-term relationship is no longer the standard by which everything else is judged.

And yet we still go in for long-term relationships, perhaps because we are reluctant to change our habits, because we are lazy or because we are afraid of the unknown. We cling to our partners because the world outside is cold and harsh. There are two other main reasons why we do so, and neither of them is likely to disappear in the foreseeable future. The first has to do with children. Once we outgrow the mad enthusiasms of our youth, the desire to have a child wells up from deep inside us and becomes irrepressible. And it is often easier to rear a child when there are two of you. Many women who have reached their thirties are interested not so much in finding a fairy-tale prince as in finding a father for the babies they plan to have. Having babies is not something to be shared by a network; it takes a well-knit team that stays together: a married couple. It is focused on one object, and the baby is the exclusive centre of attention. Even sex tends to become a secondary consideration.

The other reason is that the 'golden rule' of mutual trust and reciprocal recognition means that conjugal exclusivity works to the advantage of the individual (Kaufmann 2011 [2009]). Whatever our husband or wife does or says, they are right and their actions and words are worthy of our admiration. We are their unconditional support and absolute fan. In a world of increasingly critical and cruel judgements that can do a lot of psychological damage, such mutual trust is a source of comfort and it helps to restore our

---

[7] For an account of why the dream failed to materialize, see Kaufmann 2011 [2009].

self-esteem. Whereas the precondition for sex as leisure activity is non-commitment, long-term relationships are based upon the principle of sustainable exclusivity. Mutual support and recognition have to take priority: our partner comes first and is held in better esteem than anyone else, parents and close friends included. On the one hand, open-ended and shifting relations are subject to our desires of the moment; on the other, a closed pact can have a therapeutic effect precisely because it is closed and exclusive. This explains why it is so difficult to combine the two relational systems and why falling in love so quickly causes problems and dysfunctionality in the world where sex is merely a leisure activity. These two principles are incompatible and they are in competition.

As a rule, it is only on the margins in secrecy that these opposites can meet. In the world where sex is a leisure activity, *Missreckless* dares not admit to being in love and makes do with dreams; long-term relationships in which the partners have secret mistresses and lovers have long been the traditional staple of farce. But now that sex is openly described as a hedonistic technique and a means to well-being that is both harmless and widespread, legitimate couples want to enjoy the same right as the singles who find so many dates online. But they do so in their own way. In other words, they do all they can to respect the principles of exclusive commitment. They use a whole series of new techniques and props (books, films, sex manuals, sex toys, Viagra, and so on) in an attempt to find new ways of doing things that might otherwise become a matter of routine (Brenot 2001; Mossuz-Lavau 2002). They may test the limits of their 'exclusivity' agreement, and may even go beyond them. They may go in for an 'open' relationship (Chaumier 2004), and they may decide to experiment together. Swinging, for instance, has recently come in for some attention on the part of the media. Those involved in such practices say that they want to spice up their relationship by exploring sex as an open-ended leisure activity that can take many different forms (Welzer-Lang 2005).

## Sex is not a leisure activity like any other

If we take a closer look, it becomes apparent that it is usually men who think like this. Most women are not enthusiastic but go along with the suggestion because they do not want to disappoint or even lose their husbands. In his survey of swinging, Daniel Welzer-Lang reproduces (2005: 159) the following interview with a couple. Jean-Pierre is 37 and an engineer. Gisèle is a 39-year-old teacher. They are married with two children:

> Question: 'Were you having sexual problems?'
> J.-P.: 'I've always been a bit adventurous when it comes to sex, and it's true that it's not always easy.'
> G.:    'Yes, you really felt that something was missing.'
>
> Question: Did you talk to each other about what was missing?
> J.-P.: 'Yes.'
> G.:    'Yes, except that I found it upsetting and hurtful and always ended up in tears.'
> J.-P.: 'I think she thought that she was never enough for me, if I can put it that way.'
> G.:    'Yes, you're right. That's right.'
> J.-P.: 'That's the way I am.'
> G.:    'But, basically, what I didn't understand was what I *wasn't* doing. I had the impression that he could get it from other women but not from me. That scared me. I was frightened of being in danger, of losing him, of having to do things I didn't want to do, of letting him down, of not being good enough, and then I realized that everyone else is going through the same thing and has the same fears.'

This initial discrepancy between what men want and what women want can lead to problems. As Pascale explains (Welzer-Lang 2005: 170), 'The more I agreed to do, the more he wanted. It was never enough.' Her husband was no longer thinking about spicing up their relationship; he was thinking only of his own pleasure and was no longer concerned about what his wife wanted. When she refused to go to more than three or four 'parties' in a year, he quickly adopted the classic solution of sleeping around. Sexual liberation destroyed their relationship.

It may well no longer be as special as it used to be, but sex is not and never can be a leisure activity like any other. That can only be the case in specific contexts. Singles can enjoy sex as a leisure activity. So can couples, but not as a couple. This suggests both that the principle of non-commitment has to be respected, and that utilitarian egotism has to be kept within certain limits.

For couples who want to enjoy this kind of sex, it has to be a way of strengthening the relationship, and the desire has to be mutual.

## Sex, lies and the internet

The internet has made sexual and emotional infidelity incredibly easy. It has given married men their second wind. Men claiming to be single often use dating sites. Many of them just fantasize about the women they meet there, but some hope that it will go beyond that. An American study (Madden and Lenhart 2006) claims that 23 per cent of married men visit dating sites. The women they meet online think there are many more of them, and claim that 50 per cent of them are married. *Channelchris* (Kaufmann 2008 [2006/1999]: 124–5), who knows what she is talking about, is actually worried about the stability of their marriages and has issued a stern warning to married women:

> We've got the internet now. Even if you live in a village in the middle of nowhere, you're not safe. It's going on under your roof. There are lots of warning signs. If your husband rushes off to his computer as soon as he's finished his evening meal, ask yourself why. If he's still on his computer at two in the morning and panics when you go into the room, you've had it. If he spends whole weekends on it, you know what I think. I know what I'm talking about. I have spent two months trying out the married men who go on the net. Put yourself in his place. He chooses a nice user name, takes a few years off his age, takes up sport again, takes a photo of himself in the garden and off he goes. He has all the little chicks in the world chasing after him. Each more beautiful than the last, they all want to get to know him. And possibly more. They'll even come to the house to find him. Him. Your husband. You ought to know that he's set up a personal in-box just to meet them. Husbands who use the net are clever. So you'll never know anything about it. Unless you make the effort to catch him out.

When, in this day of sexual nomadism, people are interviewed, they are still adamant that they believe in fidelity. Fidelity is not, as we all too often tend to believe, just a moral obligation, or a survival of a long tradition that we observe just to be like everyone else. On the contrary, it is part of the pact of special recognition, which is so vital to any long-term relationship. And it is becoming more and more important. There is a growing tendency for relationships to look like a world apart and to become a psychological prop for both parties. Sex, which plays a central symbolic role in any relationship, cannot be the exception to the rule. That is why so many experiments with open sexual relationships end in failure and will go on ending in failure. Despite all the media coverage, swinging is very much a minority taste.[8]

So what do couples do when faced with these contradictory injunctions or when they are forced to choose between the attractions of sex-as-leisure activity and the promise of fidelity that is so basic to a long-term commitment? They have only two options: genuine fidelity or secret betrayal. This has of course always been the case: there is nothing new about lovers. But the internet has revolutionized the ways in which we find lovers.

It is disconcertingly easy to be unfaithful. It takes only a minimal degree of computer literacy to ask for what you want when you want it. 'Computer-assisted infidelity' is 'an emerging and growing trend' (Lardellier 2004: 147). One recent development is the appearance of sites that are specifically aimed at people who are in relationships and that offer fleeting secret liaisons. The Ashley Madison Agency, for example, uses the slogan: 'Life is short. Have an affair.' It claims to have 4.5 million members in the United States. All you have to do is to pay for a subscription that discreetly provides as many contacts as you like. The more recent Gleeden, which works in the same way, is described as the 'first international dating community dedicated to married people'. Satisfaction and discretion are guaranteed in the hypermarket of desire:

> Gleeden understands what you want, and so we have put in place a system where you only pay for what you use. There is no time limit, so you can go at your own pace: Gleeden offers you six different credit packs so that you can choose the ones that most suit your needs. You're sure to find one you like.

[8] 0.6 per cent of women and 2.2 per cent of men have had sex in a swingers' club at least once in their lives (Bozon 2008b). The percentage of people in steady relationships who have done so with their regular partners is even smaller.

Singles claim to be very interested in this expansion of the market. When it comes to sex as leisure activity, there are advantages to finding a partner who is married: because they are in a steady relationship, they are much less likely to fall in love. They really are looking for a good time and nothing else.

Online infidelity is now so popular that there are new technologies (spyware, tracking) that allow straying partners to be kept under surveillance (Chatelain and Roche 2005). There is, however, a problem: most internet affairs are short-lived and many of them are no more than one-night stands. Once again, the internet has revolutionized infidelity: lovers are beginning to look old-fashioned. Having a clandestine 'affair' used to mean leading a sort of double life and was mainly about sex. The affair was often long-lived and could involve the routines that are inherent in any long-term relationship, and the partners got on each other's nerves in the same way that married couples do. The net makes it possible to have all sorts of short flings. It also makes it look very fashionable.

## Sex/love: a historic reversal

I explained earlier how, in the course of history, sex became divorced from feeling, and gradually became autonomous before turning into a commonplace leisure activity. I then pointed out that not everything is that simple, especially when it comes to long-term commitment. We now have to look at something paradoxical.

According to the romantic ideal, it all began with sentiment, which then developed into desire. Love led (via marriage) to sex. We now seem to have two very different options: we can either cheerfully indulge in sex as a leisure activity, or we can opt for a long-term commitment. The first option means that self-control is primarily a question of avoiding commitment: we are careful not to fall (too much) in love. Hence all the subtle distinctions we have been looking at: sex on a first date means that you will never see her again; no sex means that you have feelings for her. There is a lot of confusion on dating sites: are the people who visit them looking for a 'soulmate', or just for a good time? Our thinking might be confused and inconsistent, but we are always trying to pigeon-hole things. Our expectations have to fit into one of two categories ('fun' or 'serious'), and they involve very different types of behaviour.

The strange and ironic thing about this situation is that it is often the opposite that happens: serious-mindedness gets us nowhere, but fun can. Eva Illouz (2007: 88) underlines one aspect of this paradox. Increasingly, serious dating takes the form of a process of bargaining, especially on the net. The potential dater is looking for the ideal project and trying to sell him- or herself. *Gallia* hates going on dates that have been arranged on the net because they mean that she has to do 'a sales pitch'. She feels that this encourages people to pretend to be something they are not. She goes on to explain: 'Basically, I am a very genuine person. But [when I was one of these dates] I would smile a lot, be very, very, very nice. I do not express any extreme opinion, although my opinions are extreme and I am an extremist.' She concludes: 'I do it because I really want to meet someone and because I get tired of being alone' but 'in 99 per cent of cases, I simply don't enjoy myself'. She has yet to find someone.

'Fun', in contrast, encourages people to describe themselves as they really are because it is 'not serious'. It encourages us to express our secret desires rather than retreating into our shells. It encourages us, in other words, to speak our minds. And to trust our feelings rather than coldly evaluating the ideal product. We are free to express our emotions, and that can take us much further than we expected. It also means that conjugal commitments can start with sex games in which nothing serious is at stake, or so we think. Love used to lead to sex; it is now the other way around, and sex can lead to love, even when we do not want it to because we think of it purely as a leisure activity.

So sex really is not a leisure activity like any other. It is quite impossible to divorce it completely from feeling, which always pops up where we least expect it. The idea that there might be two clearly defined categories is nothing more than an illusion. 'LoveSex' will always be ambiguous. And that does not always make life easy, especially for women.

# Part III

# Women, Sex and Love

# 9

# Unbridled Pleasure?

---

### Provisional freedom

The internet is a 'world apart', and it takes us into a completely
different parallel universe that has its own rules. All the things
that anchor our identities to reassuring but suffocating certainties
in real life suddenly vanish. Individuals have become freer than
ever before. They are autonomous and they are the equals of all
other individuals, regardless of their social class, religion or
gender. There are no (or at least many fewer) prohibitions or
taboos on the net, which makes a point of rejecting stereotypes.
This 'world apart' claims to be open-minded, free and unpreju-
diced. Women, in particular, are invited to be as open as men
about their desires.

We have already seen that things are not in fact that simple, and
there is always the danger that women will be described as 'slags'
in those parts of the net that set themselves up as the guardians of
tomorrow's morality (and especially in discussion forums). The
desire for complete autonomy and freedom represents only the
visible part of the net. There is a lot of activity in this area, and it
masks its opposite: the creation of new norms that involve harsh
judgements. And those judgements rely upon the old stereotypes.
Woman are invited to speak freely, and are then described as
'slags', as they always have been. When insults like that are being
hurled around, everything turns nasty and it is hard to understand

what is going on. Why on earth has this age-old injustice reared its head again? Why? Why can men do things that are, in theory, no longer taboo for women but that still put women at risk when they do them in real life? This is even the case on the net, which claims that men and women have the right to say everything and to ask for everything in their search for the absolute truth.

Women are not taken in by this. They hide behind their user names and counter-attack. They denounce those who slander them, often to fairly good effect (remember how *Stephie82* made *Etienne* look like a fool). The war that the 'slags' are waging for freedom and equality is being fought on the net, too. There is, however, more to this 'world apart' than online chat. There are also dates. A date is a sort of hybrid: two people meet in real life (and they probably know more about each other than people who only know each other in real life), but they are still in a 'world apart' that is far removed from the obligations and conventions that structure everyday life. In theory, the rules that are laid down in cyberspace apply to real life, too. But as soon as two people meet in a bar, things no longer work like that. As soon as bodies are involved, a long historical memory that had, we thought, been forgotten suddenly rises to the surface. And so does the rule that says that the man has to make the first move. The woman responds by sending out signals that invite him either to go on or to conclude: 'It was nice, but it was just a drink.' If this does lead, as they both hope it will, to something nicer later in bed, the woman may find that she has greater freedom to take the initiative. But at this precise moment, or at the point where someone has to make the first move, the weight of the past is such that it is hard to change things.

These oppressive stereotypes from the past come to the surface in specific contexts. In these highly coded scenes, we behave in stereotypical ways that have a long history behind them (whereas we are increasingly free to make up our own rules in other contexts). They also dominate the general schemata that define attitudes. Men and woman have taken different views of sexuality for hundreds of years. And whilst those differences are now fading, they have not really disappeared (Bozon 2008b). Men find the idea of pure pleasure with no strings attached exciting; they have long been interested in sex for the sake of sex. It is only relatively recently that women have discovered that possibility. This is quite simply because they were not allowed to think in that way: their role was to be the heart of the family, and they were expected to devote themselves to their families, body and soul (Knibielher and Fouquet 1982). When they did acquire the right to enjoy themselves, pleasure could not be divorced from love and the family. The two things went together. There were historical reasons why men could keep them separate, but that option was not open to women. The effects of this are still being felt. Women fall in love more easily than men.

## Men never change

Men are much more likely than women to be sexual predators. Their vague hints that they might be in love are designed to deceive, and they will be off once they have got what they wanted, with no or few regrets. The women they leave behind are often bitterly disappointed. *Pretty-Sophie* was delighted that she could stay the night at her Romeo's place.

> First thing in the morning, he told me 'I have to go out, I'll be an hour at most, and I'll be back by 11. I just need to see a client.' Everything was going really well, and I was over the moon: a guy who was willing to take time out from his schedule and trusted me enough to leave me in his flat. He's happy to leave me alone in his flat, so he must really like me.

*Pretty-Sophie* was quickly disillusioned:

> I decided to look at his e-mails, so I went on his computer, which was on stand by . . . And then his Outlook in-box came up on screen. I said to myself that I shouldn't read his e-mails, but my damned curiosity got the better of me. And then, to my Horror, I learned the truth. He'd had over 60 e-mails from various women. And he'd sent the same number. E-mail after e-mail to women he knew on the net. He was in touch with at least 40 of them. I was going mad and I felt sick, I closed everything down, grabbed my things and ran. I couldn't even lock the door behind me as he hadn't left me the keys.

Women have always wanted to be seduced, and they still do. They want to be told that they are beautiful, fascinating and desirable. How can they not want to be told things that improve their self-esteem? How can they fail to believe the first smooth-talking man who comes along? The bitterness is always the same, and so is the annoyance at having been taken in yet again. No two ways about it: men don't change. It is as though women's view of men was schizophrenic, as though it actually accentuated men's double standards.

Men are now changing in many respects: they take better care of their bodies, are more in touch with their feelings and are more involved with and affectionate towards their children. But in other respects they have not changed at all. They take little interest in housework (Singly 2007) and tend to divorce sex from feelings. Men are so complex and complicated that women do not have the time or patience to look at every aspect of their behaviour. So they make snap judgements. This is why they alternate between phases of naive fascination and cynicism when they come to the conclusion that men just do not change.

I noted the same ambivalence when I looked at the issue of topless sunbathing on beaches (Kaufmann 1995). Women were asked if they thought that men thought about sex when they saw half-naked women. In reply, they explained that a new code governing the way men looked at women on the beach really had changed things, or seemed to have done so. It is a refined code that has something in common with the aesthetic sublimation of the nude, the difference being that it makes semi-nudity look normal: men no longer take any notice of topless women because topless sunbathing has become so common as to be the norm. And the fact that they are not interested means that women are free to do as they please. 'So men take no notice these days? But don't they still look at you in a particular way?' As soon as I asked that question, my informants all said the same thing, as though it was obvious: 'Of course they do. Men don't change.'

*Pretty-Sophie* and many other women are constantly changing their minds about men. And it is not just that they see them differently: last night's fairy-tale prince really has turned into an ugly frog. A lover suddenly has other things to do, and even forgets his basic manners when the time comes to say that it's over. The woman concerned feels sick, disgusted and confused. Women are described as 'slags' when they talk too openly about sex on the net, but they are rejected when they show any sign of becoming emotionally attached. We still have a long way to go before we achieve equality.

## Freedom, equality and sexuality

A word about equality between men and women. Everyone, or almost everyone, now subscribes to this democratic principle, and that in itself is a step in the right direction. We have made considerable advances over the last fifty years or so, and that is as it should be. Unfortunately, there are still some major obstacles to be overcome and they do not bode well for the future (it is by no means certain that we will make further progress towards equality in the next new years).[1] The obstacles are not, however, always where we think they are. Discrimination obviously still exists in the great institutions of political and economic life, but that is nothing compared to what goes on in private life. And it is there, at the most private heart of private life, that we find the mechanisms that quietly perpetuate inequality.[2] In other books (Kaufmann 1992, 2008 (2006 [1999]), I describe how the mechanisms of married life cast men and women in complementary roles, and thus encourage women to reactivate a whole historical memory and to over-invest in their families. When it comes to sex, the mechanisms come into play much more quickly and have an even more brutal effect, even though married life and sex are obviously not unconnected (witness the fact that women become less interested in sex when they have children because of their increased family responsibilities).[3] Sexuality appears to be 'the last

---

[1] For a very complete account of both the progress that has been made and the obstacles that stand in the way of further progress, see Guionnet and Neveu 2004.

[2] Especially when they are reinforced by the demands put forward by the 'difference' feminists who have rightly been criticized by Elisabeth Badinter (2005 [2003]).

[3] The greater the division of domestic labour, the more men express desires that often remain unsatisfied.

refuge of a timeless representation of the order of things' based upon the 'naturalist illusion' (Fassin 2005: 273). This is probably the domain in which the stereotypical images of men and women are at their most powerful. Whilst tools like the internet give the impression that everything is changing very fast, women still find access to sex more difficult than men. And when they do gain access to it, they are told that they are 'slags'.

It would be a mistake to claim that this is a separate issue or a purely private matter. Being described as a slag is not just unpleasant in itself (especially as men do not come in for the same treatment). Progress towards democracy in the wider context depends upon the 'identity' models that are shaped by the subjective mechanisms that reproduce inequality. These 'cultural scenarios' have far-reaching and multiple implications.[4] In recent years, the situation has begun to change, and not necessarily for the better. For several decades, the attitudes of girls and boys seemed to be becoming similar, but there are now disturbing indications that they are beginning to diverge once more (Bajos and Bozon 2008). In the long term, this raises the possibility that the idea of sexual equality may become obsolete (Bajos and Bozon 2008: 593).When we go on the net, we see a space of freedom and equality, but this does not tell us everything about today's sexual landscape. On the contrary, some obstacles are greater than ever.

## Revolt

This is, of course, why it was this issue that provoked women into rebellion, though it was not really a central issue for the classic women's movement which was (with the exception of its lesbian-separatist elements) more at home with social and political demands than with the right to pleasure. The movement adopted the 1960s' slogan of *jouir sans entraves* ['unbridled pleasure'] and insisted that continued inequality would mean that women were not fully autonomous or independent and would lead to a democratic deficit (Bajos, Ferrand and Andro 2008). Women had to be free from

---

[4] Laumann et al. (2000) emphasize the importance of the cultural 'scripts' and 'scenarios' that have structured sexuality for years.

male sexual domination. Today's revolt may well have adopted the *jouir sans entraves* slogan, but the mood is no longer joyous and carefree. Times have changed, and a lot of promises have been broken. Hence the harsh, angry tone. Strangely enough, women's struggle for equal sexual rights leaves a bitter taste in the mouth.

It is women artists, novelists and film-makers who are putting forward these demands. Their language is often radical or extreme, and they see violence as a central part of human relationships. The demand for the right to be obscene is a response to the hatred towards women that men supposedly display (Marzano 2007). In more general terms, women journalists are popularizing the same demands in women's magazines and elsewhere. Articles entitled 'Treat Yourself to a Man Tonight' can easily coexist alongside recipes for an orange salad or articles on the latest trends in handbags. The slogans of the pioneers have become 'a groundswell: women's vision of sexuality is everywhere, on the net, in movies and in photography' (Millet 2005: 120). This media support, combined with the freedom women enjoy on the net, means that any woman can pluck up her courage and experiment with sexual liberation. 'We've decided to share our desires, fantasies and drives. With our lovers – some long term, others not so long-term.' More and more women are having one-night stands. 'Some girls behave like men, and they also go through periods of sexual bulimia with not a care for the future. It's pure greed. There are women who enjoy one-night stands, and I've met some of them' (Wingrove 2008: 189).[5] And they are happy to look for sex dates rather than wait for a man to ask them out.[6] Some, like *Woman-Warrior*, remember only the men who make them come.

Paradoxically, it is these experiments that do most to show that complete equality is a trap for the unwary. Women who become too bold quickly find that they have become the subject of some unpleasant rumours. The net is not as anonymous as we might think. Those who use it leave traces everywhere. Reputations

[5] A study of sexuality in France (Bozon 2008a), based upon a much more representative sample, indicates that there has been a significant increase in the number of young women (21.8 per cent for women aged 18–19) having relationships in which they have sex on only one occasion.

[6] *Nick* remarks in his notebooks that, in 77 per cent of cases, it was the woman who made the first move.

circulate and become inflated, and the gossip can do a lot of harm. *Lucie* is disgusted:

> There are girls who are more and more forward and independent, and who are claiming and even demanding the right to be seen as men's equals. Why should a man who has lots of women be a Don Juan and be praised to the skies, when a girl who does the same is automatically a slag? Fucking Judaeo-Christian morality. There, I've said it.[7]

And yet these new Amazons appear to have a lot going for them. Sites like Adopteunmec.com tell men like *Ginfizz* that they should become sex objects, post their profiles (proudly) and wait for a woman to click on them and put them in her online basket:

> What might make you think it's brilliant is that the guys can't hassle the ladies, and it's only after they've become an object in some woman's basket that they can send a message. That's why women who've been on other dating sites are so keen on it. So I said to myself that this is new, that for once a site shows women some respect and gives them space to go on the pull without being hassled. And for once, they're in complete control and can make the first move.[8]

Sadly, *Ginfizz* was quickly disillusioned.

## The impossible golden mean

He says he visited the site 'just out of curiosity'. He may not have been looking for a soulmate but he was looking for a bit of fun, and he was quite serious about it. We will have to take him at his word. In any case, his comments are more important than *Ginfizz* himself. He was very surprised to find that the women who had gone on sexual shopping trips alternated between two extremes: it was a case of either too much or too little. It was as though they were incapable of finding the golden mean that allowed them to

---

[7] From *Lucie*'s blog.
[8] From *Ginfizz*'s blog.

dress their sexual demands up in formulae that expressed at least a minimal degree of politeness, if not affection.

When it was a case of too little, they only *seemed* to be making the first move. They put men in their baskets, but that was as far as it went. They then waited for the man to send the first message. 'Where's the initiative in that?' asks *Ginfizz*, who finally concludes: 'Unfortunately, the ladies never make the first move. It's enough to make you weep.' When it was a case of too much, it was the way they went about things that was so depressing. They were curt and got straight to the point. They did not display any hint of humour, and not a word was said. This was pure consumerism. They were even worse than men. Given that posting a profile without a photo had produced no hits, *Ginfizz* did some work on his image. 'So after a few days, I decided to post a photo . . . but not just any old photo . . . a photo of a good-looking kid who just has to be a male model. You know the type: shirt half-unbuttoned, hair a bit messy, very *As the Bell Rings*.' The number of hits soared. There's nothing surprising about that; in fact you might say that it's only logical.

> But the worst of all is that some of them did not even look at my profile (text, description, etc). One photo, and you find yourself in her shopping basket. I don't know what to make of it all. It's like a girl seeing a guy in the street, going up to him and snogging him there and then. I find that a bit (very) crude.

*Ginfizz* is wrong to be surprised, as there is in fact nothing surprising about this kind of behaviour. These sexual scripts are the product of a very long historical process (Laumann, Gagnon, Michael and Michaels 2000; Simon 1996). They allow both men and women to see certain types of attitude as 'natural', and that makes them feel at ease. Rejecting these stereotypical attitudes feels completely unnatural. If they reject them, they lose their bearings; they have yet to develop the appropriate reflexes. That is why they alternate between too much and too little. In their 'too much' phases, men are pure predators who try to conceal their real intentions, whilst women are much more likely to adopt offbeat and provocative attitudes. Women feel they are on unfamiliar ground. Now that the safeguards created by a long cultural tradition are gone, they get drunk on their own sexual freedom and daring. And they are all the more likely to press the 'self-destruct' button in that they are to some extent aware that they are on the front line of a revolutionary struggle for freedom and equality. Sometimes, they even forget to enjoy themselves.[9]

## A cold, selfish monster

And yet the search for pleasure is the starting point for everything else. And people who are looking for pleasure can be very single-minded, even if it does mean being coldly selfish. *Saskia* is not always like this, but sometimes her needs get the better of her. She paints a sincere self-portrait:

> I'm looking for sex, and I never feel that I've had enough. I arrange dates knowing that they will be almost purely sexual. I nearly always take them back to my place. The missionary position is my favourite: I still like a good old-fashioned fuck. But I sometimes wear a blindfold so that I can indulge in fantasies that don't include the guy I'm with at the time. I get very disappointed if I don't have an orgasm; if my partner does, I demand one too, as though it was my right. I like it when he puts my needs first, as I find it easier to receive pleasure than to give it. I don't always feel like giving him a blow job, and I'm not going to force myself. The size of my partner's cock matters a lot to me. I won't try to console him if he's not up to it. If he's not a good kisser, I won't kiss him. It doesn't much bother me if a lover doesn't stay all night; sometimes it's a relief if he doesn't. I don't promise my lovers anything. I don't call them back. I wait for them to call me. I don't always plan to see them again and I tell them so to their faces. I don't ask them to be faithful, and I see no reason why I should be faithful. I've had three different partners in the space of a week, and I have seriously contemplated adultery. I've left men without saying a word, and it only took a glance to tell them that it wasn't worth them calling me again. Some men who did want to see me again got short shrift, and I've kept men hanging around for months until they got tired of waiting. Sometimes I do really fancy someone, might even become obsessed with him. But I don't fall in love.[10]

[9] Men do so too, but that is usually because they are too influenced by the spirit of competition.
[10] From Saskiablog.fr.

*Saskia* is well aware that she is sinking into a swamp of indifference when this side of her takes over and calls the tune. But what she experiences as 'sinking' is in fact moderate and discreet. It is purely subjective, and she knows what is happening to her because she begins to feel cold. When women begin to talk dirty they can, however, be spectacularly provocative. *Bagatelle*, a sexy magazine for women (it features male nudes), tries to tackle the stigma head on: 'Resolution 10: I have the right to be a slag' (Damian-Gaillard and Soulez 2001). But it is when both hunter and prey begin to role-play that things really change. The women openly go on the attack and are quite happy to be seen as predators. They are lionesses and leopardesses, and they are out for blood. A so-called trend for 'cougars' has recently emerged in Canada and the United States.[11] A cougar is a woman of between 40 and 50 who hunts very young men who are just over the age of consent; they are referred to as 'toyboys' or 'willing prey'. This role reversal obviously suggests that they are taking their revenge on men: 'Shame on me . . . but not really. This is a jungle. I used to be a gazelle, and I got eaten up. Now I'm a lioness (still in training) and I do what I like.'[12]

## More . . .

Women who are greedy for sex for its own sake and who have turned hunter may, like men, become caught up in a never-ending spiral: they want more and more because they derive less and less satisfaction from having the same experience again and again. *RedheadedGirl* is insatiable (*Nina* would describe her as a 'hysteric who's as randy as hell'). She is so quick to reject clumsy lovers, men who are not well-endowed or who come too quickly that this day's adventure is worth describing. She had set her sights on:

> a really cute guy. Fit but doesn't realize it. The kind of guy who makes you think, 'Come on, girl. You have to see what he's got inside his boxers!' It wasn't difficult. One phone call, and he was

[11] See especially CougarDate.com.
[12] Blog on *Meetic girl*.

at my place the next night. An hour after that, he was inside me. I pushed back the coffee table, knocked over a full ash tray and an empty bottle of rosé, knocked over my box of condoms, and got him on the floor. I straddled him on my knees and started to shag him. He was a bit surprised but he didn't put up any resistance.

Then we fucked. For a long time. A very long time. I came. Twice. No, three times. Or ten times. Lots of times, anyway. He didn't. It wasn't for lack of trying on my part. I tried every position I could think of. Went down on him, tried to jerk him off . . . penetration . . . kisses. I was gentle, then rough. Talked dirty, talked softly. Moaned and screamed, was submissive, then dominant. Told him to come in my face, in my mouth, on my breasts if it made it any easier. Not for an hour, not for two hours and not for three hours . . . Three and a half hours of fucking. Not a drop of sperm. Not a microlitre. In the end I had to admit defeat while I still had a vagina.

And then (and I can hear your gasps of amazement), we slept together. Yes, we did. In the same bed. Together. The two of us. As I've already said, I find it much easier to share my body than to share my bed. There's a danger you'll wake up with someone you don't really like apart from the sex. That you'll be embarrassed, that you won't know how to get rid of him. And there's a real danger that he'll be a real pain.

But *RedheadedGirl* was unsettled by the fact that he was still hard and couldn't come. She wanted to solve the problem. So she had another go first thing in the morning. Nothing doing. 'I had to leave him looking sheepish, with the erection he'd had for hours and couldn't get rid of. Fuck, shit. That really got to me.'

*Nanouhska*'s comments on *RedheadedGirl*'s blog suggest that she is not too bothered by the physiology of what happened. What she is really interested in is pick-up techniques that work: '*Nanouhska*: "So what magic spell did you use to get him to your place inside five minutes?" *RedheadedGirl*: "I said, 'Hi! It's *RedheadedGirl*. Fancy a drink tomorrow night? Come over to my place.' " '

Simple, isn't it? Although she has had her disappointments, *RedheadedGirl* is out to have a good time, and that seems to be enough for her. It is only when she is not having a good time or begins to have doubts that the self-destruct mechanism takes over. When that happens, the number of dates she has is the only thing that matters. She knows that she is going too far, and that she is probably going to get hurt. She self-harms to convince herself that she is still alive.

*Bianca* is no longer in control and has only one ambition in life: to have as many men as possible. 'That's what keeps me going: having more and more men, knowing I have good skin, and a good arse. And knowing that I can still attract a man.' She did try marriage, but the experiment was short-lived. She had withdrawal symptoms. Like a drug addict, she was dependent on her dates. She needed to have one man after another. And still she was lonely:

> The most disturbing thing about all this is that I still need to be hurt again. I'm drawn to places where I might fall into the void. I'm beginning to want it again . . . the loneliness of being the centre of attention, the wild nights, the way I'm fatally attracted to things that will hurt me. I want to hit rock-bottom. I have to be single to lead that life. I want to wallow in debauchery and excess, even if it hurts. Perhaps I'm just a little whore who's terrified of having to choose, who can't face any kind of commitment at all. Perhaps that's what all this is about.[13]

---

[13] From Bianca.hautefort.com.

# 10

# The 'Bad Boy' Paradox

### From Prince Charming to bad boy

Women, and the lives they lead today, are products of their history, and their history differs from that of men in several respects, especially where love and the family are concerned. Women and their willingness to sacrifice themselves body and soul have been crucial to the formation of family ties ever since ancient times. But when their sense of duty was in itself no longer enough to act as a categorical imperative, such commitment would have been impossible had it not been divorced from the other component of love, or the fantasy that a sentimental romanticism would enable women to escape their day-to-day existence (Kaufmann 2011 [2009]). And so, for two hundred years girls deluded themselves by dreaming of Prince Charming.

As I have found in a number of other studies, the dream persists and is still there today (even though the prince has now taken to working out in the gym to make himself more irresistible). It persists quite simply because it is too difficult to give up the dream that an emotional impulse can irresistibly lead us to what we imagine to be the truth about love. The dream is the complete antithesis of today's hypermarkets, where it becomes impossible to choose because there is so much choice. The dream also persists because we live in a society which is highly competitive and in which we are always on trial and because individuals find this mentally exhausting. We need hugs, a shoulder to cry on and the kindly looks that heighten our self-esteem. We need a place where someone who loves us can provide comfort and solace. So women make up an identikit picture of the psychological support they need: a kind man who is affectionate and who is attentive to their every need.

To be more accurate, there are moments when they make up this picture. They do so when they dream of the distant future in which they will be happy in a home of their own, with laughing children of their own and a pet of their own. At other times, or when they are thinking of a more immediate future, they develop a very different perspective. Now that sex has become a commonplace leisure activity, women want to do away with the past. There is now a radically different alternative to the past in which they sacrificed themselves for emotional reasons: they can be independent individuals who can assert that they too have the right to have a good time. All at once, the image of the ideal man changes completely. They forget about men who are kind and gentle. Sexual desire takes over. And sexual desire makes them want a 'real' man. They draw upon the most macho stereotypes to find the most exciting image possible. 'I've always found devils more attractive than angels,' says *Madamprevaricates*. *Suki* says the same thing: 'Men who are soppy, clingy and over-romantic are unbelievably tiresome and annoying.'[1]

## Why do women like bad boys?

*Nina* is very intrigued by the way both she and her girlfriends (she calls them her 'twentysomethings') find bad boys irresistibly attractive:

> Women prefer bastards. It's a fact. Do they choose them on purpose? I'm not convinced. When I look at my list of twenty-somethings, and think of Anne's relationship with her Tobias, Victoire's relationship with her Fulbert, or about Linga and her ex, I say to myself that we really do have a masochistic gene. Sorry, girls, but when I see the way men treat you, it really pisses me off.

[1] Comment on *Blonde*'s blog.

She also has another girlfriend. 'On the one hand, she's got her new official boyfriend who bombards her with loving text messages and wants to be with her all the time. On the other, there's this playboy who takes no notice of her and likes to be surrounded by pretty girls. If she was being rational about it, she'd go for the boyfriend. But she doesn't.'[2]

So why is it that even *Nina* prefers 'bad boys to nice lads'? Why did she have a scene with 'a nasty piece of work'? 'The more he treated me like a doormat, the more I grovelled. I don't understand how I could let him walk all over me like that.' In an attempt to understand, she had a talk about Ludovic with a girlfriend. Ludovic is 'a charming guy' who is 'very romantic', even 'a real sweetie', but 'there was something missing'. They talked and talked, and finally found something they could agree about. The 'but' was that he was too nice, and that was what she could not take. 'If you go out with a nice guy, you're sure to have a Prince Charming who'll do anything you ask. That's all very well, but it's no fun.' She would rather have 'someone who can talk dirty and pretend to be a woman hater, and be nice about it.'

The nice guys, for their part, are quite baffled: they just do not understand why they are so coldly rejected when all they want to do is to be kind and considerate. *Ginfizz* is sickened by women's duplicity:

> Every man has heard women talking this way. To begin with, you listen because you think they're telling the truth, because you think you'll learn the answer to the big question (or what men see as the big question): what do women want? When you first get together, you hear them say things like: 'I want a nice, loveable guy who does what I ask him to, who never loses his temper, who's kind and considerate, not a male chauvinist who'll chase anything in a skirt.' We've all heard it. And we've all been astonished and surprised to see that 'bastards' and all the guys who are a pain in the arse in the relationship department always have girlfriends and, what's even worse, succeed in picking up and 'closing the deal' with a young lady who's just said that guys like him are the last thing she wants. When he sees that, the beginner asks himself: How? Why? What spells do they cast? How do bastards like that, the ones who are such a pain, the one's who are always chasing

---

[2] *Nina*'s blog on Les Vingtenaires.

women, manage to get what they want, and why do girls go with them when they know that, once they've been dumped, they'll be telling anyone who'll listen what a bastard he was. And there was a 'nice guy' there all along.

Predatory women who are after sex and nothing else, and who want bad boys to satisfy their needs, are in the minority. But they are a very active and voluble minority, and they set the tone for the online debates. And they are a minority that is getting bigger and bigger. They have turned the world of dating upside down and triggered the emergence of a strange group that seems to have been miscast: the nice guys who have been knocked off balance by all this sexual aggression. Astonishingly, they are even more sentimental than the women. The group also includes some disappointed romantics. *Grenobleboy* was describing one of his failed dates to his friend Marie. She burst out laughing and teased him:

> Oh, come on. There's no reason why a sex date shouldn't lead to something more romantic. A sex date might well start off with a bit of slap and tickle, but that doesn't mean that it's all decided in advance and that there's no real need for anyone to seduce anyone else. I get the impression that girls want nothing to do with romantic nights out, unless they're with THE ONE. That pisses me off. I like the sideways glances, the gentle touch, the hints, beautiful views over a town that's all lit up, snowflakes falling on a pair of lovers-to-be (or a couple who are just going to spend the night together). But it seems all that's out of date.

*Nina* feels a little embarrassed when she thinks of Ludovic: 'So now, the guys will be able to say it too: "They're all bitches. They jump on you, shag you, and then they're off and you never hear from them again." Yeah, well.' She is a little embarrassed, but no more than that. She's also somewhat sceptical, as she is well aware that these frustrated romantics are not exactly thick on the ground. Like the cougars who prey on them, they stand out because they are so unusual.

At a more banal level, men may be simply surprised when a woman is bold enough to make the first move, or when she suddenly makes it obvious in physical terms that she wants sex. Even *Anadema*, a past master of the art of the pick-up, admitted to being taken aback by the cold cynicism of *Marriedwoman*, who, five minutes into their first date, 'half-opened her jacket to let me see the swell of the breast and "show me the goods", as she put it'. *Myl* was keener on showing *Ginfizz* her bum. She is very proud of her bum, and told him from the outset that her ex thought she had 'the most beautiful bum he'd ever seen'. They were dancing in a club. She was wearing skinny jeans, and she didn't miss the opportunity to show off the beautiful curves she was so proud of. So she showed off her bum, saying out loud that sex was the most important thing in life. Feeling a little embarrassed, *Ginfizz* tensed up. A disappointed *Myl* told him: 'It looks as though you've got a broom up your arse when you dance.'[3] That is as far as it went.

## In praise of pick-up artists

After many painful experiences and a great deal of thought, *Ginfizz* decided that a change of attitude was in order. 'You realize that you can't be perfect, that you mustn't try to be, that you don't have to be the nice guy.' He had to try to become something of a bad lad, too.

If nice guys are forced to reach this conclusion, what are we to make of the rest of them, of the hordes of boys who are intrinsically bad and of all the unscrupulous predators? They were beginning to feel that they were old-fashioned and felt vaguely guilty because of the way the new humanitarian and feminist values of the day made them look ... They were forced to learn to pretend and to talk the language of nice cuddles, as though they saw sex as a leisure activity, and to keep quiet about what they really wanted. And then, all at once and to their surprise and delight, women began to roll out the red carpet for them, and even encouraged them to say things they thought they'd never be able to say again. Being a bad boy had become trendy and there was no longer any need to hide. 'Everything is much simpler now that I've gone over to the light side of the force,' remarks *Macaroon*.[4]

So they are not ashamed to speak out, to start web sites dedicated to pick-up artists or to publish handbooks for the benefit of

[3] *Ginfizz*'s blog.
[4] Comment on *Blonde*'s blog.

the shy or those who are newcomers to the game. In *The Manual*, which rapidly became a best-seller amongst unapologetic pick-up artists, Steve Santagati (2007) gives practical advice on how to become the kind of bad boy that women like. Surprising as it may seem, and official values notwithstanding, the media are allowing bad boys to go legitimate, and even institutionalizing them. This is the highly paradoxical outcome of women's somewhat desperate attempt to remove the sexual obstacles that blocked the road to equality. A group of shamelessly predatory women demanded equal rights with men when it came to pleasure, and the most traditionally macho men suddenly became the cutting edge of modernity. That was all it took. The floodgates that could scarcely hold back the waters had been opened, and torrents of murky unsatisfied desires burst through them. They were crudely expressed in terms that objectified women and sometimes revealed a real contempt for them (Soral 2004). All this is dressed up in a dubious irony that makes it possible to go beyond the boundaries of what is legally acceptable.

*Ra7or* is representative of this group, and nothing, it seems, can stop him. He makes an exception for 'official girlfriends'. 'If you want to make her your official girlfriend, you should reread *Ra7or* from the beginning: sex dates just do not turn into official girl-friends. It's just not possible.' But where other women are con-cerned, his worldview is crystal-clear. 'The problem doesn't arise when you're with a sex date. What policy you adopt depends on what you expect from her. If you don't expect anything from her, don't introduce her to anyone. Just shag her and she'll be grateful.'

Personal pleasure is the only thing that counts, and there are time limits. Once the bad boy has got what he was after, he doesn't have time to talk or to get sentimental, especially as girls are stupid:

> She clears off. There's no need to be nasty, but there really are limits. You don't let a sex date stay the night. So what if she's tired and if her legs are wobbly. So they should be if you've done your job properly. I'm sure she'll sleep better at her own place. If you're so tired that you give in and let her stay, just close your eyes and imagine waking up with her. The bed is sacred; it symbolizes married life. No jokes, damn it. Yes, well, if you're shattered as

well, it might seem difficult to take a hard line. So what are you
going to do, once her ladyship has stayed the night? Sleep with her
in your arms? You must be soft in the head if you think that. She's
just a passing vagina, remember. If you want someone with arms
to hug you, you need a real chick. How many times do I have to
tell you? She can stay, but her toothbrush goes back in her handbag
straight away. No weakness. Don't give in. If you do, there's
trouble ahead: "You didn't make that clear . . . But I thought . . ."
Don't give her anything to keep her quiet. To put it in more obvious
terms: don't leave her any room for interpretation. Girls are stupid.
You say black, she hears white. You say white, she hears black.
Don't say anything!

## Return of the bastard

Women very quickly realized the scale of the disaster, without,
however, realizing that some women had inadvertently helped to
trigger it. The *connards*, as women call them, are flourishing as
never before, even though everyone thought they were a thing of
the past. Some of them are the true male chauvinists of old: they
are impenitent and have resurfaced, thanks to recent developments. Others come into different categories. Some of them are
shy and are trying to resolve their relational difficulties by pre-
tending (with some difficulty) to be ill-mannered bad boys. Some
of them are wise guys who are playing around with the bad boy
image. *Guillaume*, who blogs as *AdopteunConnard* is, to say the
least, very direct; his opening gambit is 'Hi, want to fuck?'

Some women are trying to organize a counter-attack. The Les-
bridgets.com site offers very specific advice on 'how to spot a
*connard* quickly'. But when they fall for another *connard*, their
annoyance is such that they very quickly begin to generalize and
conclude that men are all the same. As *Ginfizz* notes, some of
them were interested in sex for the sake of it but quickly changed
their minds. 'All that's over and done with. Sex for its own sake
is wrong. Men are bastards with one-track minds. They just want
to get a girl into bed and aren't interested in a serious relationship
(even though they made it clear from the start that they wanted
a purely physical relationship).' *Match* has become very touchy
about this. She sees *connards* everywhere, and she reacts all the
more violently in that she still attracts them on a regular basis

(perhaps she should ask herself why?). Like the fireman – 'good-looking but thick' – that she had a one-night stand with. The next day, he sent her a text asking if 'I *or one of my girlfriends* would be interested in a date – just for sex, no strings attached. Amazing! I remember that very clearly. They just don't give up.'[5] Then there was the 'charming young man' who was a little too crude for her liking: 'Get lost, imbecile. As though I needed a man to make me come!' Her mind is made up: 'Pigs, the lot of them.'

Remember what *Marion* had to say about FWBs, about her nice utopia and the idea that, if it was wrapped up with a little tenderness, sex could become a bubble of shared well-being, an experiment that was limited in time. If it was all about making the other person happy, it could be a romantic moment based upon physical sensations and a desire for human intimacy. It might, she hoped, offer an alternative to the competitive aggression of their normal lives. We are looking here at the opposite extreme. We are looking at a sort of hellish spiral in which the liberation of desire in fact triggers all sorts of nastiness: selfish coldness, a lack of respect, bad manners, the complete absence of even a hint of love. *Marion* was dreaming of a sort of fling with no strings attached. That is not what we are talking about here. This is no 'well-being' and the partners are not equals. On the contrary, this is a hellish spiral which exacerbates the gap that sometimes divides men from women. 'It makes everything that's bad about us worse,' complains *Saskia*. And the idea of emotional commitment becomes even more problematical. A compulsive addiction to dates that involve bodies and not souls 'becomes the best way of ensuring that you never fall in love'.[6]

Sex is really not a leisure activity like any other.

---

[5] *Match*'s blog.
[6] From *Saskia*'s blog.

# 11

# Avoiding the Traps

## Sex today

The experts who study changes in sexual behaviour have some difficulty in drawing any definite conclusions from all this. On the whole, they certainly agree that we are seeing a steady, if relatively slow, trend towards greater freedom, inventiveness and boldness, especially on the part of women: they have more partners in the course of their lifetime and are expanding their repertoire of practices. Most experts also agree that the internet is probably speeding up this process. But one cannot fail to be struck by the contrast between this seeming moderation (to say nothing of the way that so many couples fall into a quiet routine) and the unabashed exuberance that is on display on the net. I have already outlined elements of an explanation. First of all, there is the simple fact that, no matter what we might say, there is more to life than sex. To be more specific, there is an elementary need for encouragement and support. Long-term conjugal-style commitment can satisfy that need (though there is the obvious danger of lapsing into a routine or experiencing harrowing crises). We also have to conclude that it is active minorities, and not the silent majority, who set the tone. Whilst there are more women who still dream quietly of meeting their prince than there are cougars, the latter have only to roar to make some uninhibited bad boys come out of the woodwork. And if we really wish to understand how things stand, we have to recall that there is nothing fixed about behavioural categories. On the contrary, our aspirations are constantly changing and can undergo spectacular reversals. The net's cougars rarely remain cougars for the rest of their lives. At times, they may be looking for a nice husband. In most cases, the various ways in which sex and love can be articulated break down into very clear sequences.

That is why it is difficult to draw any definite conclusions. It is not just that the overall picture is very fragmentary and contradictory. We often change our opinions, and the changes can be very pronounced. This is especially true of women. When they do commit themselves to having a family, they do so wholeheartedly. When, on the other hand, the internet opens up the possibility of a sexual freedom that has nothing to do with any emotional commitment, they may experience a whole range of intoxicating sensations. They may feel that they are taking their revenge after hundreds of years of prohibition and subordination. The idea that they can drop the burden of their usual domestic commitments can make them feel light-headed, and the feeling that they can enjoy a wild time – just like men – while they are still young can be intoxicating. After all the years of suffocating conjugality, all this can feel like a breath of fresh air. And so on.

## A break from normal life

They may also want to find the strength to break some invisible chains. 'I needed at some point in my life to go through a stage that freed me from something else,' says Catherine Millet (2005: 122). This explains certain extreme or transgressive attitudes. The 'sex for the sake of it' phase may be short-lived, and may be nothing more than a sudden impulse. *Milton601* noticed that some of the women he dated in fact had very impoverished sex lives: they arranged online dates because they gave them a break from their dreary routine.[1] That is certainly not true of *Missreckless*. She does not understand what came over her last Friday night. She was in a trendy club, flung herself at a guy 'and snogged him like a sixteen-year-old'. She thought that 'such things' would never happen again, now that she had given up drinking and had,

---

[1] Discussion forum at Yahoo.com.

she said, embarked upon a one-to-one relationship (in a manner of speaking: she was involved on a regular basis with four men at the same time!). 'But nothing is normal these days, and I'm even more impulsive than ever. I spend a fortune on sexy underwear, dye my hair even more blonde, and I've begun to buy heels, even though I think I'm tall enough as it is.'[2]

These interludes can be part of a longer phase that represents a particular stage in an individual's biographical trajectory. This is often the case with young people, who can easily avoid the thorny issue of commitment for the moment. *Poupine* made the most of her youth; she had a whale of a time. There were no limits, and she has no regrets: 'There's nothing wrong about liking sex, about having a libido. Sex is not something dirty, or something to be ashamed about. The important thing is to have sex because you want to, not because you've been forced into it.'[3] She can now take a more detached view of all that frantic activity, and dreams of a very different future. 'I think it's just a phase, and that things look very different once you've settled down.' To tell the truth, she is going through a transitional period, and is not sure whether that new phase in her life still lies in the future or whether she has already turned over a new leaf. 'I think it's time to bring this phase to a close. I have done things I'm not very proud of, but there are also things that I don't regret. But actually, it doesn't matter very much, as it's all in the past.' To make it perfectly clear that she has changed, and, in an attempt to get rid of the bad reputation that clings to her, she is going to move away and start all over again. 'I've calmed down lately. I used to have a very active sex life, but not recently, and people will begin to talk . . . I think I'll move to the other side of the country in a few months and make a new start. I think I'll try to deal with things differently this time.'

### Men as sex toys

*Poupine* has turned over a new leaf without making a great fuss about it, rather as though it was natural for her life story to consist

---

[2] Blog by *Missreckless*.
[3] Auféminin.com, discussion forum.

of clearly demarcated episodes. Some women are like that. Others find this pattern very disturbing. We have to recall that this assertion of sexual freedom on the part of women is, in historical terms, very new. It is a major event and it has major implications. The bid for freedom therefore takes the form of revolt, and that can involve an uncontrollable extremism.

In that respect, the fairly extreme example of *LaNe* is interesting, and therefore worth looking at in some detail. She does not agree with *Marion*'s suggestion that there should be an emotional dimension to sex-only relationships. It would be too risky. She feels that she has more in common with an unreconstructed male chauvinist like *Ra7or*: a sex date is about sex, and that is all there is to it. And that applies to girls too:

> I'd like to make one thing clear, ladies. A sex date is a sex date. And as *Ra7or* makes clear in his article, if you forget that, one party inevitably (or almost) begins to get attached (and all too often it's the woman, given that we find it so hard to divorce sex from psychology – and therefore feelings). If he's a sex date, sex is all he's going to get. If it goes further than that, you're borderline and, even if you both know where you stand, you are putting yourself at serious risk (you'll get hurt and find it hard to get over it once it's over).[4]

Given that women are more interested than men in emotional commitment, they therefore have to be even more strictly disciplined. *LaNe* finds that the best way to avoid any emotional involvement is to think of men as sex toys. Their contribution has to be purely technical. That is the essential rule:

> A sex date has to be like a sex toy in every respect. A sex toy has to work, it has to be discreet, and it has to be there when you want it. Especially the first time. You can't let a future sex toy sweet-talk you. No, let's cut to the chase. Two consenting adults want the same thing. So say it out loud. And no pretending. If it feels right, you don't need to talk about it. You know after one night (if it takes that long!). At this stage, the most important thing is that the sex toy has to do what he's supposed to do: satisfy you. If a sex toy doesn't work the first night (or day – that's up to you),

[4] *LaNe*'s blog on Thestagedoor.fr.

there's no point in going on (let's be kind: if he really is drop-dead gorgeous, and if he was pissed that night, you might give him the benefit of the doubt and try again with him just to make sure). Don't make it too obvious that you need a little tenderness. That's not what sex toys are for. If you want sex, say so. If you are trying to compensate for something else, this is not the way to go about it. Which reminds me of *the* rule: no hugs. One sex toy told me: 'You get attached to the people you hug.' And he was right.

How can you live this life without becoming attached? By having one partner after another, of course. 'Sex toys have a limited shelf life. There comes a point when you lose interest. Having purely sexual relationships is quite difficult. You get tired in the end: even if the sex toy does become a good friend because you've spent so many nights together, there are limits.' *Saskia*, who is certainly in a position to know what she is talking about, sums it up and is a little disillusioned to find that, 'when the other becomes a sex object', something is missing. There often comes a point when even the greatest experts and the true believers realize that there are limits.

## A cycle

A fortiori, those who let themselves become involved in a physical sexuality that is devoid of all feeling are not in full control of its long-term implications. When the excitement dies down, there comes a moment when a whole range of unpleasant feelings come to the surface: they feel guilty, sated or sick. *Match*, who has, it will be recalled, reached the conclusion that 'all men are pigs', feels guilty: 'After my night with *Meetic boy*, I went home in tears and couldn't stop crying. When I was in bed with him I kept thinking of *Almostperfect*, and had to choke back my tears. God, it was sordid. I felt dirty even though I'd been willing.'[5] *Kalamity* feels that she has had enough:

Three days later, you decide it's a case of the hair of the dog that bit you, so off you go to play in the darker corners of the net. And

---

[5] *Match*'s blog.

you sign up to a dating site while you're at it. Four days later, you discover that you can still attract a man. One señor *Matanza*, one *Unfaithful*, one *Nomad* and a few others later, and you've had enough. Poor, spoiled little girl, breaking your toys one after the other.[6]

*Nina* feels sick after having shagged a few *connards*. She knew full well what they were. 'But if you're hungry, you're hungry. Even if you feel a bit dirty after shagging your *connard*, but the worst thing of all is feeling like shit, as though you just eaten a supersized Big Mac meal when you're supposed to be on a diet. It tastes good at the time, but afterwards you feel a bit ashamed.'[7] Things that were so easy and so straightforward become intolerable for the same reason. *Nina* gives the example of this exchange with *Thierry*: 'After we'd had it off, he had the nerve to contact me by text again. It went something like this. Him: "You OK?" Me: "Yes." Him: "Fancy a shag?" Me: "No, haven't the time".' *Nina* has had enough of this.

Being something of an expert on what women expect, *Ginfizz* has a cyclical theory of dating:

> After yet another break-up for various different reasons, the young lady goes back to the site. Because she wants to have her revenge and/or to make up for lost time or for some other pseudo-reason. Then she overdoses on discussions with a host of specimen males. Only one thing on her mind: she wants to be sure that she's still fanciable, that she can still get herself pulled. It's obviously not hard to get an e-mail address and to wangle a date. Most men are happy to fall for the soft words, or – even better – just don't listen and just take what's on offer. After a while – how long varies – she decides she's had enough, and when all the guys who've had a good time with her decide they'd like another roll in the hay, she blanks them. Doesn't want to know. That's all over. This is the rejection stage: they're all *connards* and so on. . . . This time, she's looking for the real thing, for the guy who wants a serious, sincere relationship. And obviously enough, he's harder to find, and then there are all the ghosts from the past who want another go (some guys never learn). After a few dates that get nowhere, she thinks she's

---

[6] *Kalamity*'s blog.
[7] *Nina*'s blog on Les Vingtenaires.

found the right guy. It may or may not last, but eventually they break up. And then it all starts again. This cycle is almost universal, and it becomes boring (and I could think of other ways of describing it). I don't understand how they can go through that cycle again and again.[8]

There is in fact an explanation. Unlike men who, for centuries, were only intermittently involved with their families, women who do commit themselves do so wholeheartedly and become the pivot around which the institution revolves. This is still true today, even though men do sometimes cook and enjoy spending time with their children. It is because the early phases of a relationship are so emotionally charged that women can abandon their old identities almost painlessly, and say farewell to the freedom of their youth. They find it much more difficult to view sex simply as a leisure activity devoid of any feelings that might lead to a long-term commitment (even though it is new and, for a while, very exciting). They can only do so for limited periods, and it is difficult to say just where they begin and where they end. To complicate things still further, the way they think about love is in constant flux: the man she initially saw as a sex object may suddenly begin to look like the man she wants to spend the rest of her life with. Women cannot indulge in sex as leisure activity with the same abandon and on the same long-term basis as some men. Even *Nina*, who is no shrinking violet, is well aware of this:

We have a distorted picture of the greedy man-eater who is just interested in sex and nothing else. Of course, there are women like that, and in some ways that can only be a good thing. It takes all sorts to make a world, and I'd just like to point out that there can be times in a woman's life when she goes in for wild sex, as well as calmer periods. But I wonder if this isn't an open invitation to be crude. Is she actually saying 'We can have a shag, go our own ways and never call each other again.' Oh come on, you also have the right to become fond of someone.[9]

[8] *Ginfizz*'s blog.
[9] *Nina*'s blog on Les Vingtenaires.

Men do not go through these different phases (which explains both *Ginfizz*'s inability to understand and his irritation). Men are heirs to an old culture of sex for the sake of it. They cling to it but, should the opportunity arise, they are willing to become sufficiently emotionally involved to commit themselves to a more long-term relationship. But this sudden change of emphasis does not impact upon a man's identity in the same way that it impacts upon a woman's. *LaNe* is trying to be *Ra7or*'s female alter ego and is trying to outdo him by singing the praises of sex in the raw. She probably won't be able to keep up with him for long.

## The dilemma

Viewing sex as a leisure activity is not the real problem. For women, it is leaving that stage behind that is the real problem. That is probably the main difference between women and men. Women find it very difficult to make the transition to the commitment phase.

This is mainly because they do not know what to think. Women know, or intuitively sense, that any emotionally based commitment is likely to turn their lives upside down (much more so than the life of a man, who will not have to take on the same domestic responsibilities). That is, in fact, personal pleasure aside, why sex for the sake of it is so exciting. And that is what they have to say goodbye to. There is also an image problem: they have left traces. All over the net. The more uninhibited the 'no strings attached' period, the worse their reputation (even though they thought they had a good reputation). She was the one who dared to make the first move. She was lively and she was full of the joys. She was the one all the guys asked out and she wasn't afraid of anything; she was a sort of higher being. As soon as commitment becomes a possibility, her reputation changes completely. Everything backfires and she has a terrible reputation. *Ani* thought she did not have this problem. She had just done 'a few stupid things' and had simply got into the habit of having sex on a first date. She did not change her habits when she met Him: 'My guy did say to me that we moved on to the serious business too quickly.'[10] Then there is

[10] Doctissimo discussion forum.

the problem of what attitude to adopt towards someone who might be a future husband and not just a lover.

So what is the solution? As we have seen, the online debates reach the curious conclusion that, if it is purely sexual, you can have sex on a first date. If you have feelings for him, it's better to wait. But not for too long. The new pattern of dating has turned everything upside down: the figure of the woman with feelings – who could be either madly romantic or quietly virtuous – has suddenly been marginalized. She has not been rejected in moral terms (the forums that are so harsh on 'slags' actually cite her as an example), but no one wants to date her in real life. Things are changing so fast that no one wants anything to do with women who say 'no'. It is as though women were in a double bind. As a disappointed *Erba* puts it: 'Either a girl sleeps around or she's seen as frigid. If she goes to bed with him too soon, she's a slag, and if she makes him wait, she's a tease; it's not easy.'[11] And as always with a double bind, there is no real solution. It is very easy for a woman to acquire a bad reputation. On the other hand, there is every reason for them to agree to play by the new rules if they still want to appeal to men.

## Why women are wallflowers

Women who refuse to join in the dance are simply ignored these days. It is no longer the plain girls who are the wallflowers, as they used to be in the old dance halls. It is the ones who stick to their principles and refuse to accept that a relationship has to start with sex. All sorts of pressures are brought to bear. The guys don't want to wait. *Fluorix* cannot stand this: 'And here's me thinking that girls had got over those stupid prejudices. We're still living in the past here.' Girlfriends are more understanding. 'There's nothing I can do about it. I'm a girl who gets a lot of attention. I think it's the *Sex and the City* effect. If you're single these days you inevitably have a lot of flings, lots of casual affairs. Sex dates, love at first sight, one-night stands...'[12] *Cheapgirl* has had enough: 'All this pressure is doing my head in.' *Nina* is well aware

---

[11] Patho108.com discussion forum.
[12] *Cheapgirl*'s blog.

that it was mainly as a result of all this pressure (mainly from boys in her case) that she began to screw around:

> Do men prefer tarts? I'm still young but, judging by my experience, they do. Well, when I was still a nice girl, not many boys took an interest in me (well, yes, they did but only the nice ones). Now that there's a whiff of scandal about me, I have to fight them off. I'm still a nice girl. I don't go out of my way to be unpleasant or to put people down, but it's my reputation for being a shagbag that attracts them . . . Probably because they fancy a night of passion with me . . . They're not interested in my conversational talents . . . Having said that, I know that men are much more interested in my bitch side than in my romantic side (if I still have one). Makes you wonder who men would rather marry: the perfect housewife or the mistress who just might to be into S&M.[13]

But many women still believe that, when they go on the Real Date, everything will change as though by magic. Their ideal man will understand if they want to wait before they have sex (and may even see it as a good sign). Sadly, he doesn't have time to wait. He gets lots of offers, and usually he can't resist them. So women have to learn how to avoid both Charybdis and Scylla. They can't be too shy if they don't want to be rejected, but they also have to avoid getting a reputation for being too forward.

Those women who are brave enough to venture onto the net in their search for love and happiness quickly learn to adapt to the new rules, even if they are not especially interested in sex for its own sake. That is the way things are these days. Women who daren't, who don't want to or who have different scenarios in mind are dismissed out of hand, and they do not understand why. And things are changing very quickly. *SingleMum* would like to stick to her principles, but she is finding it very difficult. This is the sad tale of her date with *Blue-eyed boy*. He's an old friend who recently found himself on his own again. He invited her out to a restaurant.

> It was pouring with rain when we got there. 'Wet outside, cosy inside.' And then he really begins to turn on the charm. He talks and talks, tells me jokes, pays me compliments, paints a

[13] *Nina*'s blog on Les Vingtenaires.

psychological portrait of me (and it's so accurate that it becomes annoying). Every five minutes, he tells me 'I'm having a lovely evening . . . and it's not over yet . . . really, it's not over yet.' I'm laughing inside . . . And I almost choke when he tells me that two friends can sleep together. As though it was a truth universally acknowledged. Well, no, they can't. If they do, they're not friends. Friends who have sex are lovers.

As we were talking and laughing, he took my hand and refused to let go. I'm used to the way he grabs my hand when we're playing at being an engaged couple, but he usually lets it go after a while. . . . Oh damn . . . I've not seen that look on his face before . . . What's he doing now? . . . Oh no, he's kissing me . . . Oh damn, he took me by surprise. And the worst thing of all is that it isn't unpleasant . . .

One kiss. OK. But *SingleMum* is determined that this is as far as it will go:

He won't get anything more than that tonight . . . or any other night . . . I'm building a wall – a high one – between him and me. He has to get rid of the idea that I'll throw myself at any man who pays me a little attention. Of course, it does me good to have a little tenderness, a bit of affection and a little gallantry, a peck on the cheek, an arm around me and a passionate kiss . . . but I know that none of this is real, and that our evening won't lead anywhere. We've been playing at being a couple for months now, and the only difference is that his kisses mean nothing. I'm stubborn, and he's just as stubborn, but when I say no, I mean no.

It went no further than that. And nor did their friendship. There is a price to be paid for evenings like that. It is more than likely that *Blue-eyed boy* will try his luck elsewhere.

*SingleMum* is convinced that the impulses of the moment must be held in check by an ethics of emotional sincerity. To that extent, she subscribes to the traditional morality that, ever since Plato, has tried to elevate souls. Unfortunately, that is no longer a tenable position, now that a combination of the internet and feminist demands for a right to pleasure has revolutionized the relationship between sex and love. And *SingleMum* knows it: relations between men and women are changing, and it all revolves around the issue of dating and sex. 'Why is it that every man I meet thinks he can

get me into bed just by snapping his fingers? These days, all relations between men and women seem to revolve around sex. I think
that's sad. I really do. Perhaps I'm being naive. Or idealistic. Or
perhaps I'm just being me.'

## 'LoveSex'

Everything to do with dating seems to have become simpler. It has
in fact become more complicated. Everything seems to be simple:
you just have to log on, find someone to talk to, be clear about
what you both want, and then arrange to meet for a drink. And
then go with the flow. Everything is in fact complicated; when it
comes to 'LoveSex', nothing is straightforward. A lot of people
say the opposite of what they actually mean (the predatory male
pretends to be a gentle lover,[14] and the woman conceals the fact
that she wants a steady relationship). They find it hard to tell the
difference between pleasure and feeling. It is only because the
ritual itself is so neutral and banal that we feel so calm and are
under the impression that everything is straightforward.

The fact that everything is so vague does have its advantages
– and even great advantages – given that commitment is now the
real problem. Now that supply has been extended to infinity, the
number of candidates on offer is mind-boggling. And the fact that
they are all well-informed, choosy consumers makes it even more
difficult to reach a decision. As it becomes easier and easier to
have sex, the path that leads to love becomes steeper and steeper.
We are afraid – afraid of getting it wrong.

It is in fact because commitment is becoming more and more
difficult that pleasure has become so important. Pleasure has
become an end in itself. Women are now quite happy to speak like
men and to say 'no emotional involvement' or 'there's no harm in
treating yourself to something nice'. And sometimes it does work.
Whereas the fear of getting it wrong stops us from settling for

---

[14] Drague-internet.com offers advice on how to pick up women, and suggests
ways of hiding your real intentions (sweet talk, SMSs), especially if the affair
is dragging on. 'From time to time you have to pretend you're part of a couple.'
But the aim is still the same. This advice is from a page on sex dates.

coupledom, the very idea of pleasure can get us involved with someone before we realize what is happening. The fact that every-thing is so vague means that we do not have to be frightened, that we forget about our old lives and surrender to love's adventure.

The paradoxical advantages of this set-up are, unfortunately, outweighed by its more perverse effects (and I use the term 'per-verse' advisedly). I am thinking of the way things can spiral out of control as men and women urge each other on. The more women assert their right to equality, the more men lay it on thick and come out with all the things they thought they couldn't say any more. Hence the demand for an ever-more 'liberated' sexual-ity that has in fact been 'liberated' from sentimentality and, indeed, any human consideration. We use other people and turn them into objects for our own pleasure. As *LaNe* puts it, we use them as sex toys.

As we have seen, some people are trying to resist these perverse trends. Some suggest that we can overcome the ambiguities of 'LoveSex' simply by transforming it into a stable entity. That might be one way of meeting the new and irresistible expectation that sex can become an ordinary leisure activity that has nothing to do with any long-term commitment and that allows us to expe-rience moments of pleasure and well-being in a quiet little utopia. Unfortunately, that utopia (like so many others) is likely to remain a dream that will never become a reality.

The first reason is that everyone has their own ideal rules and that no such space can emerge unless there is a minimal consensus. In *Marion*'s utopia, for instance, it is affection that matters, and it is therefore important to keep the network intact. In *Nina*'s world, in contrast, crude sex is much more important, even though there are limits:

> Each to their own. My way of life might seem a little wild (I even have sex toys,[15] so you might get the impression that I'm the ulti-mate tart), but I don't see it that way. I like to live this way, but I'm not trying to shag as many men as possible. For one thing, quantity is not the same thing as quality. I don't always keep open house. And I don't want to be looking for a new cock every week. I've better things to do with my life.

---

[15] She means human sex toys.

The second reason is that, when we arrange to meet someone for a drink, we have no way of knowing whether we are entering into a stable arrangement (about what?) or the indeterminate realm of 'LoveSex'. Dating someone is, for better or for worse, a matter of desire and feelings, and these are not things that we can control.

The real reason, however, is that we all still dream of love. And the more sex becomes just another leisure activity that has nothing to do with commitment, the harder it is to find love. Love and 'LoveSex' are an 'impossible equation' (Picq and Brenot 2009). And it is because we all still dream of love that sex can never be just another leisure activity.

# Conclusion

It all suddenly changed at the beginning of the twenty-first century. Until then, dating had followed a fairly well-defined pattern and had taken the form of a gradual process that was punctuated by various rites of passage. Young people had of course already created new spaces in which they could enjoy a certain freedom. They had begun to use flirtation and dancing to create a world of their own. And then, in the 1960s and 1970s, sexual emancipation seemed to break the last chains, but a number of traditional controls remained quietly in place. The new sexual freedom we had won made the institution of marriage less restrictive, but did not really challenge it.

As the second millennium got under way, the combination of two very different phenomena (the rise of the internet and women's assertion of their right to have a good time) suddenly accelerated this trend and transformed the landscape of dating. Sexuality had begun to be something autonomous one hundred years earlier, when it became divorced from feeling. It then broke free of all restraints. The media began to describe it in purely hedonistic terms. This created a fairly clearly defined space that had nothing to do with any commitment to marriage and that was dedicated to enjoyment. A new leisure activity had emerged. Basically, sex had become a very ordinary activity that had nothing to do with the terrible fears and thrilling transgressions of the past. There was now a vast hypermarket for love and/or sex, in which everyone was both a buyer and a seller who openly stated what they wanted and tried to satisfy their needs as efficiently as possible. All they had to do was sign up, pay a modest fee (getting a date costs less than going to see a film), or even visit a free site, write a blog or use a social networking site. Nothing could be easier.

Sex has not, however, become just another leisure activity in any established or recognized sense. Attempts are being made to establish new rules in the impassioned debates that take place on the net. These have had some success when it comes to regulating the most concrete aspects of the dating ritual, and this gives the (false) impression that this new urban game has its rules. That is far from being the case. To be more accurate, the rules relate to the formalities and the setting. They tell us nothing about the content of the game. On the contrary, the issue of sex/love is now more confused than ever. Sex has not become just another leisure activity.

This is especially true of women, who are torn between the liberating desire to remove the sexual obstacles that still stand in the path of equality, and the persistent dream of emotional commitment, which has nothing to do with sentimental mawkishness. Increasingly, feelings have subversive overtones. Received wisdom not withstanding, we have to realize that it is no longer sex that carries a whiff of scandal (except when it comes dangerously close to the limits defined by the law). Paradoxically, it is love, which seems to offer a radical alterative to the usual way of going about things. Another world is possible. And it is a world of love.

I have described this world in *The Curious History of Love* (Kaufmann 2011 [2009]). Society is not naturally focused on economics, whose current dominance is not the result of some unavoidable material development. All too often, we forget love's attempts to rule the world. Of course, they always fail, and end in confusion or worse. Unlike money, love cannot be quantified in any reliable way, and it is not an instrument of government. But its every failure has had a positive effect in the private sphere. It was in fact love that created the private sphere, divorced it from the rest of society and ensured that it obeyed completely different rules. The end result was today's schizophrenic world, in which the self is always divided.

Georg Simmel explains that society is structured by 'social forms' that teach us how to perceive the elements around us. An individual, for instance, is not just a given. There are different ways of defining what an individual is, and individuals therefore see themselves in various different ways and act in various different ways. There are now two conflicting models, and the individual is pulled in two different directions. On the one hand, there is the economic model, which assumes that individuals always act on the basis of rational self-interest. The individual is a cold, calculating creature who evaluates, compares and rules out anything that cannot satisfy his or her selfish needs. This has become the dominant model and it has taken over the world. It is now invading our private worlds, as we can see from the way we look for our soulmates on dating sites. The alternative model is supplied by love (in its various forms). This model allows the individual to abandon the egotistical self of old and to devote him- or herself to others. It is the individual – and only the individual – who creates the social bond. The feelings (which can take the form of either passionate love or humanitarianism) that lead the individual to take an interest in other people create a tiny utopia, and it offers some shelter from the cold harshness of established society.

Women are now in the forefront of this gentle revolution, which flourishes best in the context of a long-term relationship (and this is now the most difficult thing in the world). Feelings can be the start of the adventure of a lifetime: the creation of a world of one's own that does not obey the rules of selfish calculation. In what I call 'the house of minor pleasure', the ideal situation or utopia can become a reality.

This gives us a better insight into the inner duality that divides women's lives into such distinct phases. It is as though two different things were at stake. The two things are contradictory, but they are both revolutionary. The egalitarian ideal, which comes up against the stumbling block of sexuality, tells women to assert themselves as free, autonomous individuals in the same way that men assert themselves. The dream of a different world that is ruled by love, on the other hand, encourages them to forge emotional ties, even if those ties make them dependent upon men and make them take on heavy domestic responsibilities in exchange for commitment.

We were deluding ourselves when we believed, as we often did in the 1960s and 1970s, that it is sexuality and not feeling that has a revolutionary import. We believed that because sexuality had long been a threat to the established order. Sexuality is now a banal issue. Sexuality is everywhere, but it is quite in keeping with the prevailing idea that all desires can be satisfied in the vast online hypermarket where individual pleasure has nothing to do with emotional commitment. Anthony Giddens (1992) goes so far as to argue that the extreme sexual radicals are, despite themselves, counter-revolutionaries.

One could, of course, be forgiven for thinking that when sex becomes a leisure activity like any other, it can also be an alternative utopia, that it can help to create a space in which we can share our pleasure and well-being. That is what *Marion* was trying to create when she introduced emotion into her network of sex dates. We may not have heard the last of such attempts to build a world of love that could offer a radical alternative to monogamous relationships and the family as we know it today. It may still appeal to young people (things become more complicated when there are children to bring up). Two things militate against it.

The first has to do with the very notion that sex can be just another leisure activity. The idea of shared well-being does have something in common with love even though there is no question of commitment. The whole point is to give and receive pleasure, and to discover the world of the other by empathizing with it. This means that we can, at least for a while, give and receive with no thought for the future. Now we have seen that this is very much a minority view, that utilitarian and selfish behaviour is very common, and that lies, contemptuous rudeness and verbal violence are commonplace. What is worse still, a sort of hellish spiral makes men become even more macho, and then makes women respond with the same aggression.

The second factor is even more decisive. We live in a very harsh world that is based on the wretched 'calculating individual' model. Individuals are distressed, tired and have poor self-esteem. Increasingly, they feel an unsatisfied need for comfort and recognition. One night of wild passion is not enough. They need a shoulder to cry on and a sympathetic ear. They need someone to watch over them day by day, and for a long time. Only the house of minor pleasures can really provide that. And it takes commitment to build it.

The landscape of sex and love is changing very quickly before our very eyes. At the moment, a great many people want to legitimize sexual freedom and tend to see sex as the most important thing in the world. This is probably only the beginning of a real cultural revolution. But we should be under no illusion: the institution of marriage is not under threat, and nor is the dream of love. The promised revolution will be no more than a sideshow.

# Appendix: On Methodology

Researchers normally define a method for collecting data that allows it to be processed by using tried and tested techniques. They may, for instance, take representative samples or contextualize the field in which they are working. It is the researcher who chooses and organizes. That was impossible in this case. I was in a similar position to an ethnographer who is in the process of discovering a new world, the difference being that I was not dealing with a Bororo tribe but with the boundless ocean of the World Wide Web. I had to come to terms with that. I had to avoid drowning and to master the raw material I had selected as best I could. For a sociologist, the net is both a blessing and a curse. It is an invaluable resource that allows us to observe much more closely than any local survey how opinions and behaviours change. But it is also full of traps as it is difficult to verify and locate the data it supplies.

Verifying the data is, it seems to me, the lesser of the two problems. As a general rule, it is impossible to prove that a given individual is telling the truth (the same can be said of interviews or questionnaires). Unfortunately, the best way of testing the veracity of what he or she is saying is now very time-consuming. Once again, the sociologist is in a similar position to the ethnographer, who has to be patient if he wishes to get to know his informants. The sociologist must, for instance, familiarize himself with the blogs he is using rather than just picking up the odd comment. It is not always possible to do that. I chose passages at random as I surfed the net. I did not take familiarity to the point of actually contributing to my favourite blogs by posting comments. That may have been a mistake, but the techniques needed for carrying out online surveys have yet to be invented. It seemed to me that regular visits were enough to get some sense of the personality, habits and ideas of the blogger concerned, and to test his or her honesty.

One quickly gets an intuitive sense of how honest a blogger is being. Style is a particularly good indicator. The primary function of most blogs is to allow the author to arrive at a better self-understanding as others look on. The blogger becomes involved in a logic of self-revelation and that makes her become bolder and bolder, as it is the size of the audience that gives her both recognition and a sense of self-esteem, and her audience will not grow unless she is perceived as being honest. The bolder she becomes, and the more sincere she appears to be, the bigger her audience. A blogger who deceives her readers often loses her audience.

The only alternative is to develop a spectacular style and to attract an audience by being ironic, cynical or provocative. In this case, it is the form that attracts the audience and that influences the blogger, who may actually come to believe what she is saying. Only a minority of blogs come into this category, and they are not difficult to identify.

It is more difficult to check what goes on in chat rooms and discussion forums. A casual visitor can say whatever he or she thinks, and the researcher has no way of knowing whether or not it is true. The length of the discussion does, however, provide some clues. Members of discussion groups often get to know each other well, and they are not slow to point out each other's virtues and failings. This discourages impostors. *Sweet-carla* began a discussion of the classic topic of 'how soon should you have sex?':

Hi everyone! Right, well, I'm of North African descent and I recently met a boy. He's of French descent. We got on well on our first date, but after we'd talked for a couple of hours he pounced on me, started snogging me and put his arms around me. It's not that it was unpleasant but I did find it strange. So, the second time we went out, he told me he wanted to go a bit further. It was becoming obvious that he was only interested in one thing! I really like him . . . trouble is . . . I'm a virgin, I've never dared go the whole way with anyone, though I have come close to it with some of my partners. But when it came to it, I said no. So, what I'd like to know is how you go about it when you do it for the first time with someone you've met. How long before you go the whole way?[1]

*Majdoo* had scarcely had time to reply ('You think that's strange. This is 2009, my love! Personally, I expect to have sex on a first date') before *tunisonline* interrupted and put an end to the discussion: 'Just take a quick look at all the messages *Sweetcarla* or *Carlasweet* has posted. Look at all the stories she tells. She's always falling in love with someone new, and he's just a product of her imagination. Then you'll understand why everyone is pissed off with her.' *Sweetcarla*'s cover had been blown.[2]

As I mentioned earlier, the other interesting thing about chat rooms and forums is that they are classic instances of the normative production of moral and behavioural points of reference. Whilst it is extremely difficult to judge how representative a forum is, we can easily get a general idea of what makes it special and of the general nature of the group behind it. It obviously takes a great deal of time and energy to describe such things in any detail. There are, on the other hand, a number of clues (and, again, style is one) that quickly give us a general idea. It is, for example, common for different age groups to clash in an attempt to take over a forum by excluding anyone who is, in their view, too old or too young. Teenagers have their own way of conducting these debates. They tend to use short phrases and are not very interested in developing an argument. They hurl insults in an attempt to cause trouble, and their discussions can become violent. They are more interested in getting together and having a laugh than in actually trying to understand or to establish behavioural norms. 'Oldies' (those in their thirties or forties) who try to join in are pilloried. They are too 'serious', or even 'boring old farts'. The 'oldies', for their part, find the presence of teenagers annoying

---

[1] Discussion forum, Auféminin.com.
[2] This does not mean that her story is completely devoid of interest, as it tells us a lot about the way the imaginary works. It is, however, important to realize that it does not reflect her real life.

because they interrupt the search for solutions. When one group does take over a forum, the other migrates in an attempt to find somewhere quiet where they can talk. Disagreements as to which problems have to be resolved are of course legitimate, but there has to be a minimal consensus as to what is being discussed – and how it is being discussed – if there is to be an atmosphere that makes everyone feel at home.

Survey data taken from blogs and forums have both strengths and weaknesses. The weakness is that the phenomena revealed cannot really be quantified. The strength is that it allows us to get an insider's understanding of the dynamics of a process that is at the cutting edge of change. My research leads me to two very different conclusions. On the net, the world of dating has undergone a revolution; things are changing much more slowly in the 'real world', where old customs and subtle social controls still persist. On the basis of this study, it is not possible to quantify things with greater accuracy (and it would be a mistake to try to do so), to say what percentage of dates are purely sexual or what proportion of women have sex on first dates. Another type of survey might be able to tell us that. That is not what I set out to do. I set out to try to understand how the social position of sexuality has changed, the way the relationship between sex and love has been redefined, and the profound changes that have come about in the ways that people meet. And even though I cannot quantify them with any accuracy, I can say that the changes that are taking place before our very eyes are taking place very quickly: the questions raised in the conclusion may still seem marginal to some, but they may soon affect us all very directly.

To end on a more technical note: The net is unlike the world of paper books in that everything is in a state of perpetual motion. Personal blogs are especially unstable. They suddenly come to an end. Sometimes they reappear in a different form or are hosted by a different web site. This does not make the task of archiving and referencing them any easier. In order to avoid printing information that might rapidly become obsolete (this is the *habitus* of a researcher who works on paper), I have chosen to cite it and to give user names (translated into English in this edition) and simplified addresses.

I would like to thank the bloggers I felt so close to when I was carrying out this research, and I would also like to tell them how moved I am by their ability to go on telling their stories in such detail day after day, and by the way they explore their lives in such unsparing detail as they attempt to express themselves. I have often been impressed by their ability to write, to find *le mot juste* and the telling detail, and to express their emotions with such sincerity. These modern diaries contain some real gems. My congratulations to *Marion, Nina, Saskia, Channelchris, Anadema, Ginfizz* and everyone else. There is one final reason why people decide to write blogs: they want to create a work of art, using their own lives as raw material.

# References

Alberoni, Francesco (1994 [1992]), *Le Vol nuptial. L'Imaginaire amoureux des femmes*, Paris: Pocket.

Alberoni, Francesco (1999 [1997]), *Le Premier amour*, Paris: Plon.

Badinter, Elisabeth (2005 [2003]), *Dead End Feminism*, trans. Julia Borossa, Cambridge: Polity.

Bailey, Beth (1988), *From Front Porch to Back Seat: Courtship in Twentieth-Century America*, Baltimore: The Johns Hopkins University Press.

Bajos, Nathalie and Bozon, Michel (eds) (2008), *Enquête sur la sexualité en France, Pratiques, genre et santé*, Paris: La Découverte.

Bajos, Nathalie and Bozon, Michel (2008), 'Sexualité, genre et santé', in Bajos and Bozon (eds), *Enquête sur la sexualité en France*.

Bajos, Nathalie, Ferrand, Michèlle and Andro, Armelle (2008), 'La Sexualité à l'épreuve de l'égalité', in Bajos and Bozon (eds), *Enquête sur la sexualité en France*.

Bauman, Zygmunt (2003), *Liquid Love: On the Frailty of Human Bonds*, Cambridge: Polity.

Belcher, James (2006), *Online Dating: Whose Space?* New York: E-Marketer.

Beldjerd, Sofian (2008), *Goûts en mouvements: L'Individu à l'épreuve d'une sociologie des activités esthétiques ordinaires dans les espaces du quotidien*, sociology thesis, Université Paris Descartes.

Beltzer, Nathalie, Bajos, Nathalie and Laporte, Anne (2008), 'Sexualité, genre et conditions de vie', in Bajos and Bozon (eds), *Enquête sur la sexualité en France*.

Beltzer, Nathalie and Bozon, Michel (2008), 'Les Séparations et leurs suites: rencontres sexuelles et prevention après une rupture conjugale ou amoureuse', in Bajos and Bozon (eds), *Enquête sur la sexualité en France*.

Bologne, Jean-Claude (1998), *Histoire du sentiment amoureux*, Paris: Flammarion.

Bozon, Michel (2008a), 'Premier Rapport sexuel, première relation: des passages attendus', in Bajos and Bozon (eds), *Enquête sur la sexualité en France*.

Bozon, Michel (2008b), 'Pratiques et rencontres sexuelles: un repertoire qui s'élargit', in Bajos and Bozon (eds), *Enquête sur la sexualité en France*.

Bozon, Michel and Héran, François (2006), *La Formation du couple*, Paris: La Découverte.

Brenot, Philippe (2001), *Inventer le couple*, Paris: Odile Jacob.

Brym, Robert J. and Lenton, Rhonda L. (2001), *Love Online: A Report on Digital Dating in Canada*, Toronto: MSM.

Chatelain, Yannick and Roche, Loïc (2005), *In Bed with the Web: Internet et le nouvel adultère*, Saint Quentin en Yvelines: Chiron.

Chaumier, Serge (2004), *L'Amour fissional. Le Nouvel Art d'aimer*, Paris: Fayard.

Clair, Isabelle (2008), *Les Jeunes et l'amour dans les cités*, Paris: Armand Colin.

Corbin, Alain (1990 [1987]), 'The Secret of the Individual', in Michelle Perrott (ed.), *A History of Private Life: Vol. 4. From the Fires of Revolution to the Great War*, trans. Arthur Goldhammer, Cambridge MA: Belknap Press of Harvard University Press.

Courier, Paul-Louis (1820), 'Pétition à la Chambre des députés pour les villageois que l'on empêche de danser', *Oeuvres complètes*, Paris: Bibliothèque de la Pléiade.

Damian-Gaillard, Béatrice, and Soulez, Guillaume (2001), 'L'Alcôve et la coquette. Presse féminine et sexualité: l'expérience éphémère de Bagatelle (1993–4), *Réseaux*, 19(105): 101–59.

Dumesnil, Annette (2004), 'Mes Copains sur Internet, c'est "pour de faux" et "pour de vrai"', *La Lettre de l'enfance et de l'adolescence* 55: 47–52.

Duret, Pascal (2009), *Sociologie de la compétition*, Paris: Armand Colin.

Edwards, Logan (2008), *Secrets of the A Game: How to Meet and Attract Women Anywhere, Anyplace, Anytime*, Los Angeles: Sweetleaf Publishing.

Fass, Paula (1977), *The Damned and the Beautiful: American Youth in the 1920s*, New York: Oxford University Press.

Fassin, Eric (2005), 'Démocratie sexuelle', *Comprendre* 6: 263–76.

Fein, Ellen and Schneider, Sherrie (2002), *The Rules for Online Dating: Capturing the Heart of Mr Right in Cyberspace*, New York: Pocket Books.

Ferrand, Michèlle, Bajos, Nathalie and Andro, Armelle (2008), 'Accords et désaccords: variations autour du désir', in Bajos and Bozon, *Enquête sur la sexualité en France*.

Fourier, Charles (1967 [1816]), *Le Nouveau Monde amoureux*, Paris: La Presse du réel.

Giami, Alain (2005), 'Santé sexuelle: la médicalisation de la sexualité et du bien-être', *Comprendre* 6: 97–115.

Giddens, Anthony (1992), *The Transformation of Intimacy: Sexuality, Love and Eroticism in Modern Societies*, Cambridge: Polity.

Guillebaud, Jean-Claude (1998), *La Tyrannie du plaisir*, Paris: Seuil.

Guionnet, Christine and Neveu, Eric (2004), *Féminins/masculins. Sociologie du genre*, Paris: Armand Colin.

Heshmat, Dina (2006), 'Le Caire traque ses amants', *La Pensée du Midi* 27: 16–22.

Hess, Rémi (1989), *La Valse. Révolution du couple en Europe*, Paris: Métailié.

Illouz, Eva (2006), 'Réseaux amoureux sur Internet', *Réseaux* 138: 269–72.

Illouz, Eva (2007), *Cold Intimacies: The Making of Emotional Capitalism*, Cambridge: Polity.

Jacotot, Sophie (2008), 'Genre et danses nouvelles en France dans l'entre-deux-guerres: Transgressions ou crise des représentations?', *Clio. Histoire, femmes and sociétés* 27: 225–40.

Kaufmann, Jean-Claude (1992), *La Trame conjugale. Analyse du couple par son linge*, Paris: Nathan.

Kaufmann, Jean-Claude (1995), *Corps de femmes, regards d'hommes. Sociologie des seins nus*, Paris: Nathan.

Kaufmann, Jean-Claude (2002), *Premier Matin. Comment naît une histoire d'Amour*, Paris: Armand Colin.

Kaufmann, Jean-Claude (2004), *L'Invention de soi*, Paris: Armand Colin.

Kaufmann, Jean-Claude (2008) [2006/1999], *The Single Woman and the Fairy-Tale Prince*, trans. David Macey, Cambridge: Polity.

Kaufmann, Jean-Claude (2011 [2009]), *The Curious History of Love*, trans. David Macey, Cambridge: Polity.

Kinsey, Alfred, Pomeroy, Wardell and Martin, Clyde (1948), *Sexual Behaviour in the Human Male*, Philadelphia and London: Saunders Company.

Knibielher, Yvonne and Fouquet, Catherine (1982), *Histoire des mères, du Moyen Age à nos jours*, Paris: Hachette-Pluriel.

Lagrange, Hugues (2003), *Les Adolescents, le sexe, l'amour*, Paris: Pocket.

Lardellier, Pascal (2004), *Le Coeur net. Célibat et amours sur le Web*, Paris: Belin.

Laumann, Edward, Gagnon, John, Michael, Robert and Michaels, Stuart (2000), *The Social Organization of Sexuality: Sexual Practices in the United States*, Chicago: The University of Chicago Press.

Lejealle, Catherine (2008), 'La Difficulté de construire son identité numérique dans la rencontre amoureuse en ligne', *Consommations & Sociétés* 8.

Lipovetsky, Gilles (2005), 'Orgie hard, sexe sage', *Comprendre* 6.

Madden, Mary and Lenhart, Amanda (2006), *Online Dating*, Washington: Pew Internet and American Life Project.

Marcuse, Herbert (1955), *Eros and Civilization*, Boston: Beacon Press.

Marzano, Michela (2007), *La Pornographie ou l'épuisement du désir*, Paris: Hachette-Pluriel.

Masson, Christelle (2006), *Journal de mes rencontres sur Internet*, Paris: Thélès Editions.

Merkle, Erich and Richardson, Rhonda (2000), 'Digital Dating and Virtual Relations: Conceptualizing Computer-Mediated Romantic Relationships', *Family Relations* 49(2): 187–92.

Miller, Daniel and Slater, Don (2000), *The Internet: An Ethnographic Approach*, Oxford: Berg Publishers.

Millet, Catherine (2005), 'La Liberté sexuelle', *Comprendre* 6.

Modell, John (1991), *Into One's Own: From Youth to Adulthood in the United States, 1920–1975*, Berkeley and Los Angeles: University of California Press.

Mossuz-Lavau, Janine (2002), *Le Vie sexuelle en France*, Paris: La Martinière.

Mystery (2007), *The Mystery Method: How to Get Beautiful Women into Bed*, New York: St Martin's Press.

Online Publishers Association (2005), *Online Paid Content US Market Spending Report*, New York: OPA.

Pearl Kaya, Laura (2009), 'Dating in a Sexually Segregated Society: Embodied Practices of Online Romance in Irbid, Jordan', *Anthropological Quarterly* 82(1): 251–78.

Picq, Pascal and Brenot, Philippe (2009), *Le Sexe, l'homme et l'évolution*, Paris: Odile Jacob.

Reich, Wilhelm (1951 [1936]), *The Sexual Revolution*, London: Vision Press.

Robinson, Paul (1976), *The Modernization of Sex: Havelock Ellis, Alfred Kinsey, William Masters and Virginia Masters*, New York: Harper & Row.

Santagati, Steve (2007), *The Manual: A True Bad Boy Explains How Men Think, Date and Mate – and What Women Can Do to Come Out on Top*, New York: Crown Publishers.

Segalen, Martine (2009), *Rites et rituels contemporains*, Paris: Armand Colin.

Simon, William (1996), *Postmodern Sexualities*, London: Routledge.

Singly, François de (2005), *L'Individualisme est un humanisme*, La Tour d'Aigue: Editions de l'Aubre.

Singly, François de (ed.) (2007), *L'Injustice ménagère*, Paris: Armand Colin.

Sohn, Anne-Marie (1993 [1992]), 'Between the Wars in France and England', trans. Arthur Goldhammer, in François Thébaud (ed.), *A History of Women in the West. Vol. 5: Toward a Cultural Identity in the Twentieth Century*, Cambridge MA: Belknap Press of Harvard University Press, pp. 92–119.

Sohn, Anne-Marie (1996), *Du Premier Baiser à l'alcôve*, Paris: Aubier.

Sohn, Anne-Marie (2003), 'Les "Relations filles-garçons": du chaperonage à la mixité (1870–1970)', *Travail, genre et sociétés* 9: 91–109.

Soral, Alain (2004), *Sociologie du dragueur*, Paris: Bibliothèque Blanche.

Strauss, Neil (2005) *The Game: Penetrating the Secret Society of Pickup Artists*, New York: Regan Books.

Strauss, Neil (2007), *The Rules of the Game*, New York: HarperCollins Publishers.

Welzer-Lang, Daniel (2005), *La Planète échangiste. Les Sexualités collectives en France*, Paris: Payot & Rivages.

Wheeler, Deborah L. (2006), *The Internet in the Middle East*, Albany: State University of New York Press.

Wingrove, Lewis (2008), *Des Souris et un homme. Un An de rencontres sexy sur le Net*, Paris: Pocket.